Elementary Lessons in Voltaic Electricity

by W. Jerome Harrison

with an introduction by Roger Chambers

This work contains material that was originally published in 1895.

This publication was created and published for the public benefit, utilizing public funding and is within the Public Domain.

This edition is reprinted for educational purposes and in accordance with all applicable Federal Laws.

Introduction Copyright 2018 by Roger Chambers

Self Reliance Books

Get more historic titles on animal and stock breeding, gardening and old fashioned skills by visiting us at:

http://selfreliancebooks.blogspot.com/

Introduction

I am pleased to present yet another title in our "How To ..." series.

The work is in the Public Domain and is re-printed here in accordance with Federal Laws.

As with all reprinted books of this age that are intended to perfectly reproduce the original edition, considerable pains and effort had to be undertaken to correct fading and sometimes outright damage to existing proofs of this title. At times, this task is quite monumental, requiring an almost total "rebuilding" of some pages from digital proofs of multiple copies. Despite this, imperfections still sometimes exist in the final proof and may detract from the visual appearance of the text.

I hope you enjoy reading this book as much as I enjoyed making it available to readers again.

Roger Chambers

PREFACE.

THIS little book is intended strictly as an *introduction* to the science of which it treats. The Education Department requires that the teaching of Elementary Science in the schools under its direction should be "purely descriptive and experimental," and I have here attempted to indicate how such teaching should be carried on. Our *theories* may and do frequently change, but the *facts* of Nature upon which they are founded are immutable. Let us study the facts, and the theories will adjust themselves.

Each and every chapter of this book has been given as an object-lesson many times to classes of children averaging sixty in number, and of ages from ten to sixteen. Every encouragement should be given to young students to experiment on their own account; and it will be found that they are wonderfully eager to do so, and also that many of them are very apt in the construction of simple apparatus. Care has been taken to describe mostly such experiments as any student may repeat at home, with but little expenditure of money.

W. J. H.

BIRMINGHAM, *January 1895.*

CONTENTS.

I.	NATURE OF VOLTAIC ELECTRICITY: THE CONTACT THEORY,	7
II.	THE SIMPLE VOLTAIC CELL,	15
III.	CONNECTION OF CHEMISTRY AND ELECTRICITY,	20
IV.	CHEMICAL ACTION IN THE CELL,	25
V.	POLARIZATION IN THE CELL,	29
VI.	TWO-FLUID CELLS,	33
VII.	THE FORCES BY WHICH VOLTAIC CURRENTS ARE URGED OR RETARDED,	38
VIII.	CHEMICAL ACTION OUTSIDE THE BATTERY,	46
IX.	HISTORY OF CURRENT ELECTRICITY,	55
X.	MAGNETIC EFFECTS OF THE VOLTAIC CURRENT,	63
XI.	ELECTRO-MAGNETS,	69
XII.	THE ELECTRIC TELEGRAPH,	74
XIII.	HEATING EFFECTS OF THE CURRENT,	80
XIV.	LUMINOUS EFFECTS OF THE CURRENT,	85
XV.	INDUCED CURRENTS,	89
XVI.	THE INDUCTION COIL,	96

APPENDIX.

EXAMINATION QUESTIONS,	101
QUESTIONS SET BY H.M. INSPECTORS OF SCHOOLS,	104
APPARATUS REQUIRED FOR EXPERIMENTS IN VOLTAIC ELECTRICITY,	105

VOLTAIC ELECTRICITY.

I.—NATURE OF VOLTAIC ELECTRICITY: THE CONTACT THEORY.

1. Names employed, and their Origin—2. Electricity as a Fluid—3. The Two-Fluid Theory of Electricity—4. Signs employed to represent the Two Electrical Fluids—5. Electricity produced by Contact—6. Electrical Condition of Metals immersed in Liquids—7. Connection between Frictional and Voltaic Electricity.

1. Names employed, and their Origin.—The kind of electricity whose effects we shall describe in this book is known by several names. The method of producing electricity by the rubbing together of two substances—*frictional* electricity as it is called—goes back more than two thousand years; but the discovery of *voltaic* electricity was made only a single century ago.

(1.) VOLTAIC ELECTRICITY is so named after Professor Volta of Pavia in Italy (died 1826), whose discoveries, made public about the year 1800, laid the true foundation of the science. It is, however, also known as—

(2.) GALVANISM, or GALVANIC ELECTRICITY, after Professor Galvani of Bologna, who made the original observations by which Volta's attention was drawn to the subject.

(3.) CURRENT ELECTRICITY is another name for voltaic electricity, because this kind of electricity flows, or appears to flow, as a continuous stream or current from one place to another when we afford it an opportunity of doing so.

(4.) LOW TENSION ELECTRICITY as a name expresses the fact that this kind of electricity is unable to overcome resistances in the same manner as the "high tension" electricity produced by friction.

(5.) DYNAMIC ELECTRICITY (from the Greek word *dunamis*, power) as a name implies the fact that an electric current has *power* to do work. It can exert force, and is therefore able to overcome resistance.

(6.) The term CONTACT ELECTRICITY refers to Volta's theory, that the electric current was produced by the simple contact or touching of two dissimilar metals; while the name—

(7.) CHEMICAL ELECTRICITY points to the opposite theory, supported by Faraday, that the production of the electricity is not due merely to the contact of the metals, but that it is owing to the chemical action which accompanies the contact.

On the whole, the names "Voltaic Electricity" and "Current Electricity" are most frequently employed for the special force whose effects we have now to consider, and whose laws of action we must try to understand.

2. Electricity as a Fluid.—In considering electrical phenomena, it is convenient to *conceive* of electricity as a *fluid*—that is, as something which can *flow* from place to place. We know that the term "fluid" includes gases as well as liquids; and in order to be able to represent the facts to our minds, it is useful to consider electricity as a very rare fluid—something of the same nature, for example, as the fluid called the *ether*, which we believe to fill all the space between the Earth and the Sun, Moon, and stars. The fact is, that we do not know precisely what electricity is. A recent theory considers electrical force as due to a vibration or to-and-fro motion of the smallest parts (molecules) of which matter is composed; but, on the whole, it is at present most convenient to retain the older theory, which considers electricity to be a fluid.

3. The Two-Fluid Theory of Electricity.—As there are certainly *two kinds* of electricity, we must not think of it as one fluid, but as *two*. These two electrical fluids are unlike one another; indeed, they are exactly opposite in all their properties. The names of *positive* and of *negative* have been given to the two electrical fluids respectively. The two-fluid theory of electricity states—

(1.) That every substance in an ordinary (or neutral) state contains an equal and unlimited amount both of positive and of negative electricity.

(2.) That when these fluids are *separated*, as by friction, by chemical action, or by any other means, so that there is a greater quantity of the one fluid than of the other upon any substance, then that substance shows signs of electricity, or is "electrified."

(3.) That "like fluids repel, while unlike fluids attract, one another." Positive repels positive, but attracts negative; while negative repels negative, but attracts positive.

4. Signs employed to represent the Two Electrical Fluids.—It is convenient to represent positive electricity by the sign + (plus), and negative by the sign − (minus). By *positive* electricity we understand that which is of the same kind as the electricity produced upon glass by friction with silk; while the type of *negative* electricity is the kind which we find on sealing-wax after it has been rubbed with flannel.

5. Electricity produced by Contact.—When two dissimilar metals are made to touch each other, one of them becomes positively and the other negatively electrified. To prove this, let a short rod of copper, c, be soldered to a similar rod of zinc, z. There will then be found an excess of + upon the zinc and of − upon the copper. To show this, hold the rod by its zinc end and touch the lower copper plate of a condensing electroscope (Fig. 1) with the copper end of the rod, placing a finger at the same time upon the upper copper plate of the electroscope. Now remove first the finger and then the rod, and raise the upper or condensing plate; the gold leaves will then diverge with − electricity, which must have passed to the electroscope from the copper. This appears to prove that when the metals zinc and copper are brought into contact, the former becomes positively and the latter negatively electrified. This is Volta's celebrated "contact theory;" but other great electricians believe that the electricity produced is due to chemical

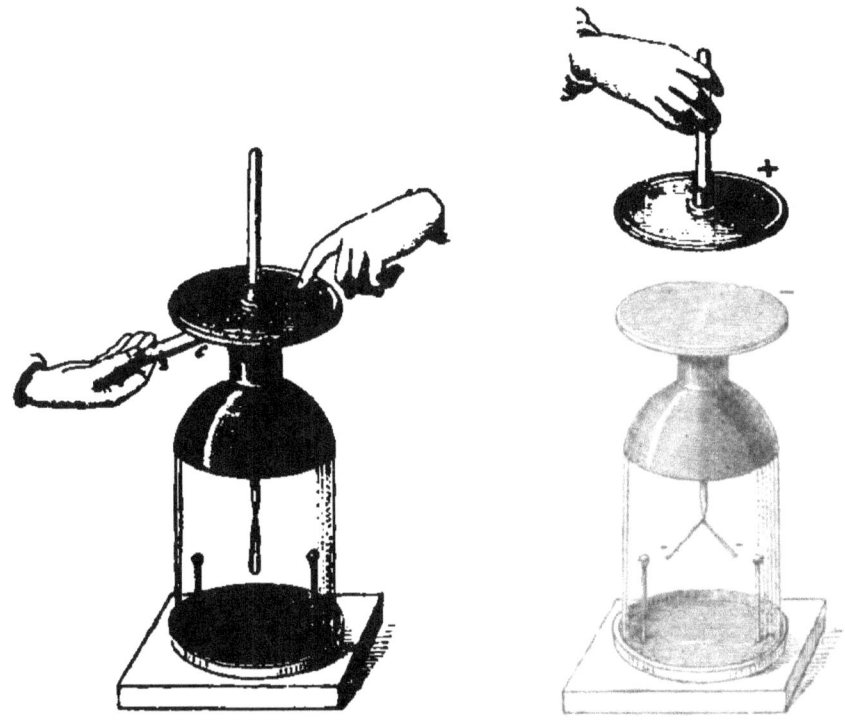

Fig. 1.—Compound copper and zinc rod with condensing electroscope.

action taking place between the metal and the moisture of the hand. We may perhaps consider that the force required to press the two metals together produces a *separation* of their electricities; but that to *maintain* a current of electricity from the one metal to the other, chemical action is necessary. In this way the contact theory of electricity may be reconciled with the chemical theory.

6. Electrical Condition of Metals immersed in Liquids.—Let the glass jar, $a\,b$ (Fig. 2), be half-filled with water, to which a little sulphuric acid has been added; place a strip of copper and another of pure zinc (each strip having a copper wire soldered to it) in the acid liquid. If the wires are now carefully tested, the one fastened to the zinc will be found to be negatively electrified, while that connected with the copper will show positive electricity.

The testing of the electrical condition of these wires is a rather difficult experiment.

(1.) With a sufficiently delicate electroscope (such as the quadrant electrometer invented by Lord Kelvin) the matter is easy enough. By touching the wires in turn with a proof-plane, we can remove sufficient + from the wire connected with the copper, and sufficient − from the wire joined to the zinc, to prove that the two wires are oppositely electrified as long as the metals stand facing one another in the acid water.

(2.) Another method is to connect the ends of the two wires with the insulated brass plates, $c\,d$, of a condenser (Fig. 2), the brass plates being each connected also with a delicate gold-leaf electroscope. Join c and d for a moment by some conductor, as the finger and thumb of one hand, or a bent piece of metal; then remove the two wires; lastly, separate the brass plates. The gold leaves of the electroscope which was connected with the condensing-plate c will now diverge, being charged with negative electricity which came from the wire fastened to the strip of zinc; the leaves of the electroscope connected with the plate d will diverge with positive electricity which had its origin in the strip of copper. For this reason the wire fastened to the zinc is called the *negative pole*, while that joined to the copper is named the *positive pole*.

7. Connection between Frictional and Voltaic Elec-

tricity.—Whether we produce electricity by friction, or by contact, or by chemical action, it is the same force—the same two electrical fluids are obtained.

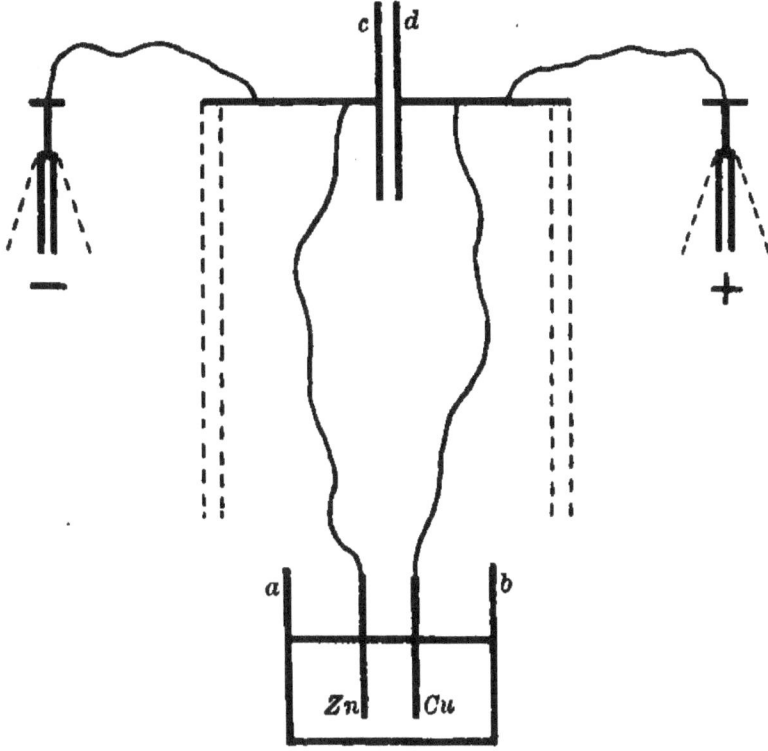

FIG. 2.—Apparatus to show electrical condition of metals immersed in liquid.

(1.) Thus in the experiment described in the last paragraph we can prove that it is negative electricity which has passed from the zinc pole to the electroscope, by bringing excited sealing-wax (the type of −) near to the disc of that electroscope, and obtaining *increased* divergence of the gold leaves. In the same manner excited glass held near the other electroscope proves that positive electricity has passed to that instrument from the copper pole; for the approach of the positively charged glass causes the gold leaves to move still further apart, and we know that "like electricities repel."

(2.) Another proof, both of the fact that electricity is produced by contact and that it is of the same double nature as that produced by friction, may be obtained by performing the experiment represented in Fig. 3. Here z and c are semicircular pieces of copper and zinc; a +ly charged very light pointer or index, usually composed of aluminium, is suspended so as to hang just above and parallel to the two metals. When the two metals are brought together the pointer *moves towards the copper*, showing that the latter has received a negative charge. If the pointer be charged −ly, it will move towards the zinc. After a few minutes the pointer returns to its old position, for in that time the two separated electricities slowly combine, and each piece of metal is again restored to the neutral state.

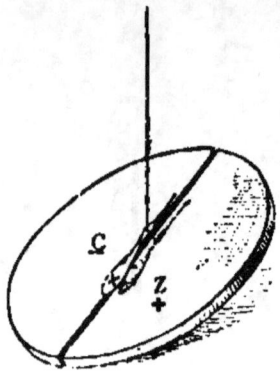

Fig. 3.—Magnetic needle over two semicircles of copper and zinc.

(3.) In the same way, when the prime conductor and the rubber of a frictional electrical machine are connected by wires with the binding-screws of a galvanometer (see Fig. 28), a deflection of the magnetic needle is produced when the machine is worked, similar to that effected by a voltaic current.

II.—THE SIMPLE VOLTAIC CELL.

8. Contact of Similar Substances—9. Local Action—10. Amalgamation of Zinc—11. The Simple Voltaic Cell—12. Action of the Simple Cell—13. Acidulated Water.

8. Contact of Similar Substances.—It will be found that, to produce electricity, the metals brought into contact with one another, or dipped into any liquid, must be *different* in some way or other. If two precisely similar pieces of the *same* pure metal, as two strips of zinc or two strips of copper, be immersed in acid water, *no* electricity is generated upon them, whether they touch or whether they are kept apart.

9. Local Action.—Acid water produces no effect upon *pure* zinc; but pure zinc would be very expensive, and the zinc which is employed for electrical purposes is not pure, but contains small quantities of other metals, principally iron and arsenic. Moreover, the surface of ordinary zinc is not of a uniform nature, some parts being harder and of different texture from the adjoining parts, owing to the operations of annealing and rolling the zinc plates. The consequence is that when a piece of common zinc is immersed in acid water it begins to dissolve, and bubbles of a gas (hydrogen) are seen to rise through the water. The reason of this is that every particle of iron or of arsenic on the surface of the zinc causes a little current of electricity to flow between it and the adjoining zinc; other currents, moreover, pass from the softer to the harder portions of the zinc. This is known as *local action*; and it is injurious, because the zinc is used up in producing electricity which remains *within* the vessel containing the liquid, while we want the electricity to pass *out of* the vessel and along the wires which are used to connect the metal plates.

10. Amalgamation of Zinc.—There is a very simple way of causing commercial zinc to act as if it were pure zinc. Dip the zinc plate into acid water and leave it there for two or three minutes. Pour some mercury into a saucer, and tie a piece of flannel, or a little tow, to the end of a short piece of wood. Now take out the zinc and rub the mercury all over it with the flannel. The mercury combines with the outer layer of the zinc, forming zinc amalgam, thus separating the zinc from any other metals which may be present, and, moreover, producing a shining coating of a pure and *uniform* nature all over the zinc plate. The mercury will not combine with the particles of iron, arsenic, etc., as it does with the zinc, so that these metals are detached or separated from the zinc amalgam, which covers them over so that they do not come into contact with the exciting liquid contained in the cell. This process is called amalgamating the zinc. Amalgamated zinc acts just like pure zinc, and when it is placed in acid water no local action takes place. In the experiment which we shall describe, we will assume that the zinc employed has been properly amalgamated.

11. The Simple Voltaic Cell.—To the pair of dissimilar metals described in paragraph 6 the name of a *galvanic pair* is given; but when we include also the liquid in which the metals are immersed, and the vessel containing it, the whole is known as a *simple voltaic cell*.

Hitherto the electricity which we have found to be produced upon the surfaces of metals in contact, or facing one another in a liquid, has been similar in its behaviour to that developed by friction. But frictional electricity either (1) remains upon the surface on which it is produced, or (2) passes away from that surface with such rapidity that the current in which it flows is only of momentary duration.

For example, the two coatings of a charged Leyden jar are in different electrical states, the inner coating being charged with, say, + electricity, and the outer coating with the − fluid. When the two coatings are connected by a conductor, the + flows through the metal in an outward current, while the − forms a current flowing *into* the jar. But these currents only last for a very small fraction of a second, because in that time sufficient electricity passes in each direction to reduce each of the coatings to a neutral state. But if we could *maintain a difference* between the two coatings, keeping the inner one always charged positively, and the outer one negatively, then a *continuous* current of electricity would flow along the metallic conductor by which the coatings are connected.

In a Leyden jar it is not possible to *keep up* such a difference when the coatings are united by a conductor, but with a *simple voltaic cell* such continually flowing currents can be developed and maintained with ease.

12. Action of the Simple Cell.—Take a strip of zinc and a strip of copper and place them facing one another, an inch or two apart, in some acid water, so that they are half covered by the liquid (Fig. 4). No visible change of any kind takes

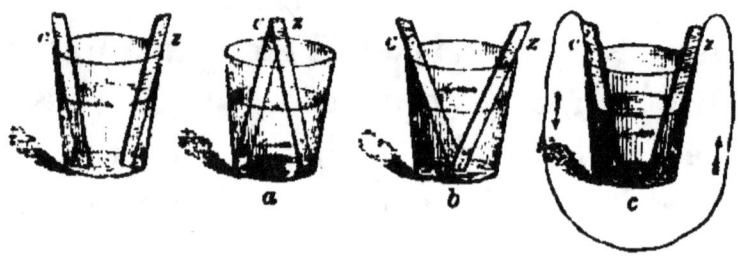

Fig. 4.—Simple Cell. *a*, Metals touching above liquid; *b*, Metals touching below liquid; *c*, Metals connected by wire.

place. Now incline the strips so that they touch *outside* the liquid. Immediately the zinc begins slowly to dissolve, while bubbles of hydrogen gas rise from the copper plate (Fig. 4). Reverse the metals, so that they touch *below* the surface of the liquid, and precisely the same action is observed. Or the metals may be immersed completely in the acid water, and the same phenomena will occur when they are made to touch. It will be noticed that only *one side* of each metal plate is of use— namely, the sides which face one another.

Instead of allowing the upper end of the copper

THE SIMPLE VOLTAIC CELL. 19

to touch the upper end of the zinc, it will make no difference in the working of the cell if we connect them by any conductor, as a piece of copper wire. As conductors of electricity, the metals stand far above all other substances; while such things as glass, sealing-wax, and gases will not allow electric currents to flow along them. When the connecting wire is properly tested, a current of + electricity is found to flow along it from the copper pole to the zinc; while an equal current of − electricity flows along the wire at the same time, but in the contrary direction, from the zinc pole to the copper. In speaking of *the* current, however, we always mean the *positive* fluid, neglecting, for convenience, the existence of the negative current altogether.

13. Acidulated Water.—Pure water is rendered a much better conductor of electricity by the addition of a little acid. But in the simple cell the presence of an acid is required for another purpose of which we shall speak later on. The proportions in which the two liquids should be mixed vary from 1 (of acid) to 8 (of water) to 1 to 12. If we measure twelve pints of water into an earthenware pan, and then add gradually one pint of strong sulphuric acid, we shall have a mixture which (after stirring it well and allowing it to cool) we may place in large bottles and label "battery acid," and which will be found very useful in voltaic electricity. By using the larger proportion of acid (1 to 8) a stronger current of electricity will be obtained, but the zinc will be dissolved much more rapidly.

III.—CONNECTION OF CHEMISTRY AND ELECTRICITY.

14. Proofs of the Existence of a Voltaic Current—15. Chemical Composition of Matter—16. Composition of Compound Substances—17. Bodies, Particles, and Molecules—18. Decomposition by Electricity.

14. Proofs of the Existence of a Voltaic Current.—The existence of the current of electricity in the wire joining the two metals of a galvanic pair cannot be directly ascertained by means of the senses. We cannot see or feel the electric fluid; but we can detect its presence by its *effects*—by the changes which it produces in the condition of the wire. If the wire be very thin and short, the passage of the current will make it hot, perhaps even red-hot. But the simplest proof of the passage of a current along a wire is the effect which the wire then produces upon a balanced magnetized needle. Separate the wire from the simple cell, and it produces no effect upon the magnet; but connect the wire with the metals in the cell, so that the electricity can flow, and then hold the wire near to and parallel with the magnet, and the poles of the latter will instantly move away from the wire (Fig. 5). By this simple method it is always easy to tell whether or not a current is flowing along a conductor; and the stronger

CONNECTION OF CHEMISTRY AND ELECTRICITY. 21

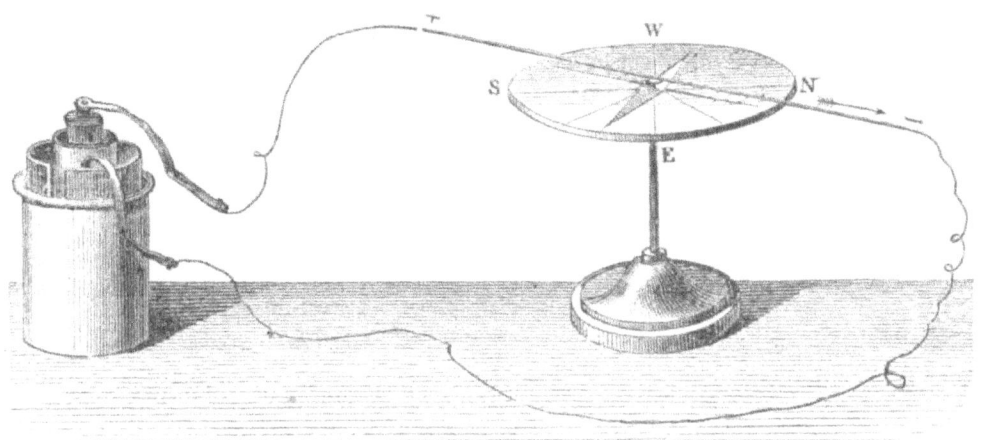

FIG. 5.—Needle deflected by current in wire placed over it.

the current the greater will be the motion or deflection of the magnetic needle.

15. Chemical Composition of Matter.—So far as we know, this world only contains about seventy simple substances or ELEMENTS, each made of *one* kind of matter only, and each different from all the others. Of these seventy elements, fifty-five are metals and fifteen non-metals. For convenience' sake, instead of writing the full name of an element, we usually employ a *symbol* for it, consisting of the first, or the first and second letters of its name. The metals most commonly employed in electricity are:—

Name.	Symbol.	Name.	Symbol.
Copper	*Cu.*	Potassium	K.
Zinc	Zn.	Sodium	Na.
Platinum	Pt.	Chromium	Cr.

The non-metals chiefly used are:—

Name.	Symbol.	Name.	Symbol.
Hydrogen	H.	Nitrogen	N.
Oxygen	O.	Sulphur	S.
Chlorine	Cl.	Carbon	C.

16. Composition of Compound Substances.—Each

symbol stands for a single *atom* of the element it represents.

Of the thousands of different substances which we see every day, by far the greater number are COMPOUNDS, being made of two, three, or more elements chemically combined with one another. Thus, water is a compound of hydrogen and oxygen; glass is a compound of silicon, sodium, and oxygen; and so on. We represent the composition of any compound by placing together the symbols of the respective elements of which it is composed; thus, for sulphuric acid we write H_2SO_4, because each molecule of that liquid is composed of two atoms of hydrogen united with one atom of sulphur and with four atoms of oxygen.

The principal compounds employed in voltaic electricity are:—

Water	$\overset{+}{H}_2\overset{-}{O}$.
Hydrochloric Acid	$\overset{+}{H}\overset{-}{Cl}$.
Sulphuric Acid	$\overset{+}{H}_2\overset{-}{S}O_4$.
Common Salt	$\overset{+}{Na}\overset{-}{Cl}$.
Sulphate of Zinc	$\overset{+}{Zn}\overset{-}{S}O_4$.
Bichromate of Potash	$\overset{+}{K}_2Cr_2\overset{-}{O}_7$.
Oxide of Zinc	$\overset{+}{Zn}\overset{-}{O}$.

17. Bodies, Particles, and Molecules.—Each group of symbols stands for one *molecule* of the compound which it represents. A molecule is the smallest portion of any substance which can have a separate existence. Thus a block of rock-salt may be called a *body*. The block may be pounded in a mortar, and

CONNECTION OF CHEMISTRY AND ELECTRICITY. 23

we then obtain a white powder composed of *particles* of salt. If this powder is shaken up in water it will disappear, for the water separates the *molecules* of which each particle is composed one from the other, and when they are alone or separate the molecules are too small to be visible, even through the most powerful microscope. Chemists have proved that each molecule of common salt is composed of one atom of a metal called sodium, combined with one atom of the gas known as chlorine. Now we may conceive that these two atoms are held together (to form the molecule NaCl) by the force of electricity, the metal being positively and the gas negatively electrified. It is the same with other compounds, and in the list given above we have marked over the symbol for each atom in each compound the kind of electricity with which it is believed to be charged.

18. Decomposition by Electricity.—If we consider

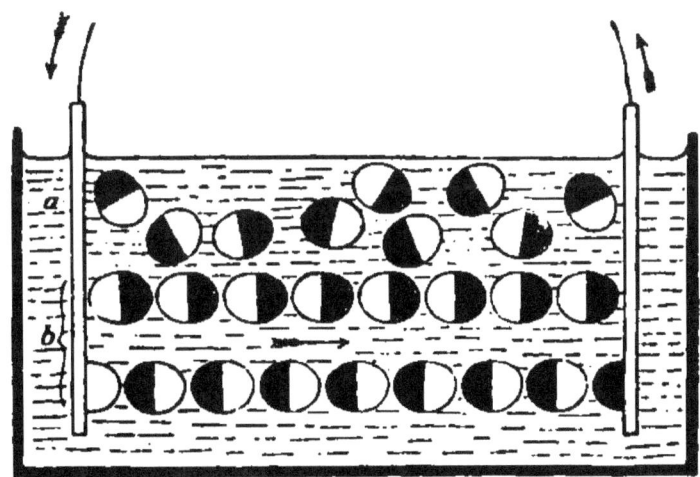

Fig. 6.—Water molecules (*a*) before and (*b*) after decomposition.

that in a molecule of water the two hydrogen atoms are charged positively, while the oxygen atom is charged negatively, it will help us to understand how these two elements combine to form water. But it is also clear that if we place some other element more strongly electrified than either of the atoms in question within the water, it may decompose or break up the molecules of water. If we place *two* elements in the water—one electrified +ly and the other -ly—then the atoms forming the water molecules will be drawn in opposite directions, and may even be dragged asunder (Fig. 6).

From these considerations it is plain that there is a close connection between the science of chemistry and the science of electricity. A knowledge of electricity will therefore be useful to every chemist, while some knowledge of chemistry is required by every student of electricity.

IV.—CHEMICAL ACTION IN THE CELL.

19. Chemical Action in the Simple Cell—20. What becomes of the Hydrogen?—21. What becomes of the Oxygen?—22. Action of the Acid—23. Direction of the Current—24. Production of the Electric Current.

19. Chemical Action in the Simple Cell.—When the plates of zinc and copper are placed facing one another in water, the zinc below the surface becomes electro-positive and the copper electro-negative. Then any molecule of water lying between the two metal plates will be attracted in opposite directions, the electro-positive atoms of hydrogen in the molecule having a tendency to go to the negative copper, while the electro-negative atom of oxygen endeavours to proceed towards the positive zinc. We must carefully distinguish between a *plate* and a *pole*. The zinc is the positive *plate*, while it is the negative *pole*. On the other hand, the copper constitutes the *negative* plate but the *positive* pole.

When the two metals are connected by a conductor (as a copper wire), they become so strongly charged that the molecules of water lying between them are actually decomposed—that is, the oxygen is parted from the hydrogen. This change may be briefly expressed by the following chemical *equation*:—

$$H_2O = H_2 + O$$

One molecule of water *yields* one molecule of hydrogen *and* one atom of oxygen.

Such a statement as this is called by chemists an equation, because the number of atoms on each side must be *equal*. In the above example it will be noticed that there are *three* atoms on each side.

20. What becomes of the Hydrogen?—Let us first follow the positively-electrified hydrogen atoms as they pass towards the negatively-electrified copper plate. When they reach its surface no *chemical* combination takes place, but the hydrogen gas rises in little bubbles from the surface of the copper—briskly at first, but afterwards more slowly—until at last the copper is covered with a film, or thin coating, of the hydrogen gas.

21. What becomes of the Oxygen?—The −ly-charged oxygen soon arrives at the +ly-charged surface of the zinc plate, but, watch closely as we may, no bubbles of oxygen gas are to be seen there. The fact is, that the zinc and the oxygen have a strong liking—or *affinity*, as chemists term it—for each other, and they therefore combine to form oxide of zinc:—

Zn	+	O	=	ZnO
One atom of zinc	*combines with*	one atom of oxygen	*to form*	one molecule of oxide of zinc.

22. Action of the Acid.—Oxide of zinc is a white solid, insoluble in water. If this compound were permitted to remain upon the zinc, it would soon form a hard crust all over its surface, which would effectually stop the production of electricity. But as soon as it is formed it is removed from the zinc by the sulphuric acid, and a clean surface of zinc is thus left ready to attract more oxygen. The acid and the

CHEMICAL ACTION IN THE CELL.

oxide combine to form sulphate of zinc, which *is soluble* in water. We may represent the action thus:—

ZnO	$+$	H_2SO_4	$=$	$ZnSO_4$	$+$	H_2O
Insoluble oxide of zinc	*combines with*	sulphuric acid	*to form*	soluble sulphate of zinc	*and*	water.

23. Direction of the Current.—The current of positive electricity is considered as starting from the surface of the positive plate—that is, from the part of the zinc which is *below* the liquid. It proceeds through the liquid to the copper plate, up the copper, and along the copper wire to the zinc again. The entire path which the electric current describes is called its *circuit;* and when the copper and the zinc are joined or connected by the wire above and by the liquid below, so that the current can flow round freely, the circuit is said to be *closed.* When the wire is cut, or removed from either of the plates, or if one metal is lifted out of the liquid, the circuit is said to be *broken,* or *interrupted,* for it is then impossible for the electricity to flow.

The ends of the metal plates which stand out of the liquid in the cell, or the ends of the wires connected with those plates, are known as the *poles,* or *electrodes,* of the cell or battery of cells.

24. Production of the Electric Current.—Of the two metals which we use to form the simple cell, one must be acted on by the acid liquid more than the other. The current produced is proportional to the chemical action which goes on at the surface of the zinc, and is maintained by it. If a strip of gold and a strip of platinum be used instead of the plates of zinc and copper, we can get *no* current of electricity, for the acid water cannot act chemically upon either of these metals. In an ordinary simple cell the zinc is continually wasting away, as may be proved by weighing it every two or three hours, while the copper plate undergoes no such loss of weight. As the cell continues working, more and more sulphate of zinc is formed; and if the liquid is allowed to evaporate, white feathery crystals of this substance will be seen upon the sides of the vessel. We thus see that the production of electricity depends upon the chemical action which takes place in the cell. When this action stops, the current ceases to flow; while, as the chemical action increases, the strength of the current increases also. The simple cell, in fact, is an example of the conversion of chemical energy into electrical energy.

V.—POLARIZATION IN THE CELL.

25. Cause of Polarization—26. Why Polarization diminishes the Strength of the Current—27. How to diminish Polarization—28. Smee's Cell—29. The Bichromate Cell.

25. Cause of Polarization.—Let the poles of a simple cell be joined by a short, rather thick copper wire which passes over and is parallel to a balanced magnetic needle. The poles of the magnet move away from the wire through a certain angle, which may be measured by placing a card, with degrees marked upon it, beneath the needle. Let us suppose that the deflection of the needle is at first 15°; then, after the current has been flowing in the wire for ten or fifteen minutes, the angle will probably be only 8° or 10°, and in half-an-hour it may be reduced to 3° or 4°, or even less. This diminution of the angle between the needle and the wire means that the electric current decreases in strength as the time for which it has been flowing increases. What can be the cause of this falling off? If we carefully examine the copper plate of the simple cell, we shall see that the hydrogen bubbles, which when the poles are first connected rise freely, adhere more and more to the copper, until at last they form a film or covering over its entire surface. When this has been accomplished, the copper plate is said to be *polarized*.

26. Why Polarization diminishes the Strength of the Current.—Gases are bad conductors of electricity, and the hydrogen gas coating the copper offers a *resistance* to the current of positive electricity which is trying to reach the copper plate. The consequence of this is that *less* electricity is able to pass along the plate to the wire joining the poles, so that the current in the wire is diminished, and is therefore less able to act upon the magnetic needle.

Another reason is that hydrogen is more electro-positive than copper, so that there is less electrical difference between hydrogen and zinc than between copper and zinc. When the copper is coated with hydrogen, it acts as if it were a plate of hydrogen instead of a plate of copper. If the two plates in the simple cell are *alike*, they produce *no* current. Now the more nearly they are alike in their electrical states, the *less* will be the current which passes between them.

27. How to diminish Polarization.—If the copper plate is lifted out of the cell for a moment, the hydrogen will escape from its surface, and the cell will then again yield a strong current. Or if air is forced through the cell, or if the negative plate is occasionally *brushed*, the polarization will be, for the moment, destroyed. It will be observed that these methods effect only a *temporary* cure, and are not therefore quite satisfactory.

28. Smee's Cell.—In the year 1840 the late A. Smee constructed a cell in which the negative plate was made of platinized silver, the positive plate being as usual of zinc. The deposit of platinum with which the silver plate is coated has thousands of small sharp points upon it (Fig. 7), and the bubbles of hydrogen cannot adhere to such a rough surface so easily as they could to the smooth surface of metallic copper. Thus with Smee's cell the current does not decrease in strength so quickly as with the ordinary simple cell. Still the remedy is not a perfect one; it only *delays* the polarization, it does not *prevent* it.

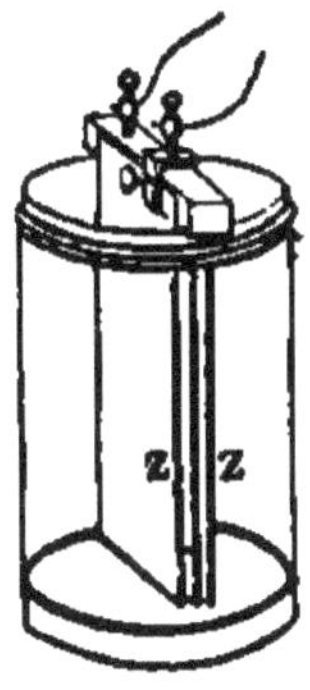

FIG. 7. Smee's Cell.

Another point we may notice in Smee's cell is that the plate of zinc is double, one zinc being placed on each side of the platinized silver. As the two zincs are connected by a metal screw at the top they act like one plate, but in this way *each side* of the silver plate is used, and a current of double strength is therefore obtained.

29. The Bichromate Cell.—The best " one-fluid " cell consists of two plates of carbon and one of zinc immersed in acid water saturated with bichromate of potash. The hard gas-carbon obtained from the retorts used in the manufacture of coal-gas is sawn into thin plates, and has the advantage that it is not acted upon by any acid (Fig. 8). The zinc plate is fastened to a rod, so that it can be drawn out of the liquid when a current is not required. The bichromate of potash combines with the hydrogen as soon as it is

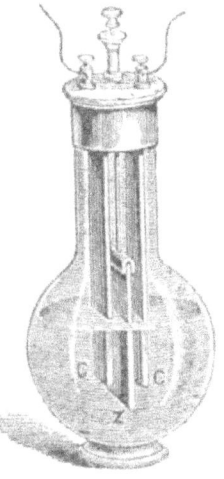

FIG. 8.—Bichromate Cell

liberated, and so prevents it from ever reaching the negative plate. In this cell, then, polarization is prevented by chemical means. For ordinary experiments the bichromate of potash cell is very useful, since it gives a fairly strong and (for an hour or so) constant current, and produces no disagreeable fumes.

VI.—TWO-FLUID CELLS.

30. Total Prevention of Polarization—31. Chemical Action in Daniell's Cell—32. Construction of Grove's Cell—33. Chemical Action in Grove's Cell—34. Bunsen's Cell.

30. Total Prevention of Polarization.—About the year 1836, Professor Daniell of King's College, London, discovered a means whereby the accumulation of hydrogen upon the copper plate could be altogether prevented, at least for a time which could be measured by months instead of hours.

Daniell's cell differs from those already described in that it contains *two* distinct fluids, which are separated from each other by a partition of unglazed baked clay, known as a *porous cell*. Such a partition prevents the free mixture of the liquids in a cell, while it allows the electric current to pass through its pores. The ordinary form of Daniell's cell has a porous cell containing acid water and a rod of zinc. This porous cell is placed inside a copper vessel, and the space between the two is filled with a saturated solution of sulphate of copper (Fig. 9). To keep the solution

FIG. 9.—Daniell's Cell.

saturated, some crystals of copper sulphate are placed in a little copper-wire basket, which is attached to the inside of the copper rim. Thus, as in the simple cell, the copper vessel forms the negative plate (and positive pole), while the rod of zinc in the centre of the cell constitutes the positive plate (and negative pole) (Fig. 10).

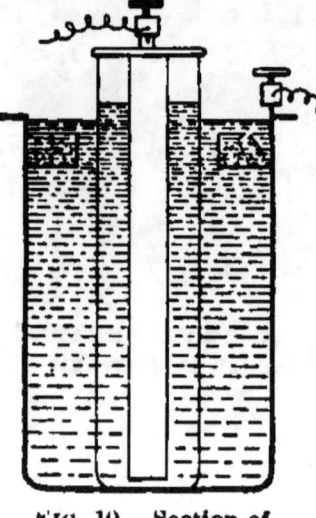

FIG. 10.—Section of Daniell's Cell.

31. Chemical Action in Daniell's Cell.—The chemical changes which take place in a Daniell's cell are as follows:—

(1.) The water is decomposed by the attraction of the two opposite electricities situated upon the two metal plates of the cell.

$$\underset{\text{Water}}{H_2O} \underset{\text{is decomposed into}}{=} \underset{\text{hydrogen}}{H_2} \underset{\text{and}}{+} \underset{\text{oxygen.}}{O}$$

(2.) The oxygen goes to the zinc plate and combines with the metal.

$$\underset{\text{Oxygen}}{O} \underset{\text{and}}{+} \underset{\text{zinc}}{Zn} \underset{\text{produce}}{=} \underset{\text{oxide of zinc.}}{ZnO}$$

(3.) The oxide of zinc is removed by the sulphuric acid, which combines with it to form sulphate of zinc, a substance soluble in water.

$$\underset{\substack{\text{Oxide of} \\ \text{zinc}}}{ZnO} \underset{\substack{\text{combines} \\ \text{with}}}{+} \underset{\substack{\text{sulphuric} \\ \text{acid}}}{H_2SO_4} \underset{\substack{\text{and} \\ \text{forms}}}{=} \underset{\substack{\text{sulphate} \\ \text{of zinc}}}{ZnSO_4} \underset{\text{and}}{+} \underset{\text{water.}}{H_2O}$$

So far, the equations show that the chemical actions which occur within Daniell's cell are similar to those in the simple cell—the water is decomposed and the oxygen got rid of in exactly the same way.

But when we follow the hydrogen, which seeks to pass to the copper plate, we find that matters are so arranged that this gas is intercepted, so that it never reaches the copper, and cannot, therefore, polarize it.

(4.) The hydrogen of the decomposed water passes through the pores of the porous cell on its way to the copper plate. But as soon as it gets outside the porous cell, the hydrogen comes into contact with the solution of sulphate of copper. Then we get the following chemical change:—

$$H_2 + CuSO_4 = H_2SO_4 + Cu$$
Hydrogen *combines with* sulphate of copper *to form* sulphuric acid *and* copper.

The metallic copper thus liberated is passed on to the copper plate, on the surface of which it is deposited, so that this plate is constantly increasing in thickness.

Owing to the constancy of its action, and to the equal and steady current of electricity which this cell consequently maintains in any wire joining its two poles, Daniell's cell is much used for telegraphic purposes.

32. Construction of Grove's Cell.—After Daniell had shown that it was possible to prevent polarization by surrounding the negative plate with some liquid that would combine, chemically, with the hydrogen liberated from the water, Mr. (Justice) Grove saw that nitric acid would answer the purpose even more effectively than sulphate of copper. But it would not be possible to employ *copper* for the negative plate in that case, for the nitric acid would attack and dissolve the copper. Grove therefore replaced the copper by *platinum* (which is not affected by any single acid), and devised the cell shown in Fig. 11.

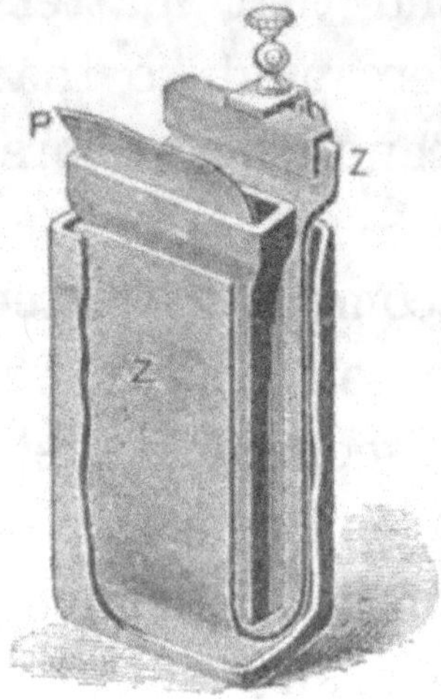

In Grove's cell a thin sheet of platinum is placed in a narrow, flat-sided, porous cell, filled with strong nitric acid. Outside this porous cell is a sheet of zinc (bent into the shape of the letter U), immersed in acid water. The outer vessel,

FIG. 11.—Grove's Cell.

which contains the whole, is made either of porcelain or of ebonite.

33. Chemical Action in Grove's Cell.—The oxygen is disposed of in exactly the same way as in the simple cell and in Daniell's cell. The hydrogen passes inwards through the porous vessel, endeavouring to reach the (negative) platinum plate. But as soon as the hydrogen touches the nitric acid, the latter parts with an atom of its oxygen, and this oxygen atom combines with the hydrogen to form water.

H	+	HNO_3	=	H_2O	+	NO_2
Hydrogen	*combines with*	nitric acid	*to form*	water	*and*	nitric peroxide.

One drawback to the use of Grove's cell is the escape of red, suffocating fumes of nitric peroxide gas. If more than three or four of these cells are used at once, they should be kept outside the lecture-room, and the wires made sufficiently long to reach from them to the table. Grove's cell produces a strong current, and is, on the whole, the one most used for ordinary experiments in voltaic electricity.

34. Bunsen's Cell.—The platinum used in Grove's cell is an expensive metal, costing about fifty shillings per ounce, so that Grove's cells (pint size) cost from nine to twelve shillings each. To make a cheaper cell, a German chemist, named Bunsen, used carbon instead of platinum to form the negative plate, and this form is consequently known as Bunsen's cell (see Fig. 12). It consists of a square rod of gas-carbon placed inside a circular porous cell filled with strong nitric acid. The porous cell is placed within a stout, circular, earthenware pot, which contains a cylinder of zinc, and is nearly filled with acid water (one part of sulphuric acid to ten parts of water). The chemical action which produces the electric current is exactly the same as that in Grove's cell.

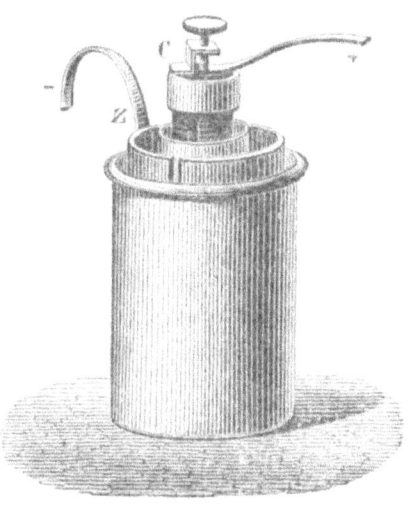

FIG. 12.—Bunsen's Cell.

VII.—THE FORCES BY WHICH VOLTAIC CURRENTS ARE URGED OR RETARDED.

35. Potential—36. Electro-motive Series—37. Resistance—38. Laws of Electrical Resistance—39. Electro-motive Force of Voltaic Batteries—40. Advantage of Coupling Cells in Series—41. Reduction of Internal Resistance of a Voltaic Battery—42. Advantage of Coupling Cells for "Quantity."

35. Potential.—We use the word *potential* to signify the relative degree to which a body is charged with electricity. As the standard of comparison, we take the electricity with which the surface of the Earth is always charged, and which is said to be of *zero potential*. Bodies electrified +ly are said to have a *high potential*, while those electrified −ly are at a *low potential*. It is plain that an electric current will always flow from places of *high* to places of *low* potential, when they are joined by a conductor. The force urging the current to flow from one metal plate to the other, in the different voltaic cells which we have described, will clearly depend upon the *difference* of potential of the two metals. This force is known as the electromotive force; it is briefly written E.M.F. If any two metals are electrified to an *equal* degree, then *no* current will flow when they are connected. But if one metal be more strongly *charged* than the other, then, when they are joined by a wire, a current

FORCES AFFECTING VOLTAIC CURRENTS. 39

will flow from the metal whose potential is higher to the metal whose potential is lower. It must not be forgotten that when we speak of *the* current, we refer to the flow of *positive* electricity. We may compare difference of potential to difference of *level;* for just as water tends to flow from a higher to a lower level, so does electricity endeavour to flow from bodies whose potential is high to those with a lower potential.

36. Electro-motive Series.—Volta discovered that different metals were electrified to different degrees when brought into contact with one another. Thus the difference of potential between zinc and iron is less than that between zinc and copper. In the following list the most electro-positive substance is placed first, and the most electro-negative last :—

1. Zinc.$^{+}$	4. Iron.	7. Gold.
2. Tin.	5. Copper.	8. Platinum.
3. Lead.	6. Silver.	9. Carbon.

Thus, if plates of zinc and tin be connected by a wire and placed in acid water, a *feeble* current will flow from the zinc through the liquid to the tin. If the tin be now replaced by a plate of copper of equal size, a stronger current will be obtained; while plates of zinc and carbon will give the strongest current of all. In constructing a cell, it will clearly be advantageous to select substances as *far apart* as possible on the list. This is one reason why a Grove's cell gives a *stronger* current than a Daniell's. The difference of potential between zinc and platinum is greater (and therefore the

E.M.F. is greater) than that between zinc and copper.

37. Resistance.—There is no *perfect* conductor of electricity; all substances offer more or less *resistance* to the passage of the electric current along them. This resistance is far more important in the case of a voltaic current than with the intense charges of electricity produced by friction. Thus the difference of potential between the two coatings of a charged Leyden jar is thousands of times greater than the difference between the two plates of a simple cell. But the E.M.F. is proportional to the difference of potential, and therefore the current from a Leyden jar is able to *overcome* a resistance which would be an effectual bar to the passage of the current produced by a simple cell.

The metals are by far the best conductors of electricity, but they differ among themselves in their conducting powers. If the conductivity of silver be represented by 100, then we have copper 98, brass 22, platinum 18, mercury $1\frac{1}{2}$. Compared with the metals, all liquids are far worse conductors; while gases are almost perfect non-conductors. Pure water is almost a non-conductor for the low-tension currents produced by chemical action; its conductivity is much improved by the addition of acids, or even of common salt.

38. Laws of Electrical Resistance.—Taking the case of the metals, it is found that their resistance depends on—(1) the *length* of the conductor: fifty miles of copper wire offer just fifty times the resistance of one mile. (2.) On the *diameter* of the conductor: a rod one inch thick will conduct four times better than a rod only half-an-inch thick; if one wire is three times as thick as another, it will offer only one-ninth of the resistance; if it is four times as thick, its resistance will only be one-sixteenth, and so on. (3.) Lastly, the resistance depends on the *material* of which the conductor is made: taking wires of silver and of platinum of equal lengths and diameters, the resistance of the platinum will be five-and-a-half times greater than that of the silver. The practical unit of electrical resistance is called the *ohm*, and is the resistance offered to the flow of an electric current by a standard piece of thin copper wire (about number 40 of the wire gauge) $18\frac{1}{2}$ inches in length.

39. Electro-motive Force of Voltaic Batteries.—The current produced by any *single* cell—whether it be Smee's, Daniell's, or even Grove's—cannot have much electro-motive force. If we merely increase the *size* of the two plates of any cell, we do not increase the electro-motive force; for the *difference* of potential between two square *inches* of any two metals is the same as that between two square *feet* (or any other area) of the same metals. For this reason a single cell, no matter how *large* it may be, is unable to decompose water; it has not enough E.M.F. The atoms of oxygen and hydrogen which form each molecule of water hold together with a certain force; to separate them we must employ a *greater* force. Now there is no single cell which has a force greater than that by which the atoms in a molecule of water are held together. But by joining cells together *in series*, connecting the positive metal of the first cell with the negative metal of the second, and so on, we can form voltaic *batteries*, whose electro-motive force will be just as much greater than the force of one cell as the total number of cells composing the battery. Thus, in Fig. 13, let z be the zinc plate of the first cell.

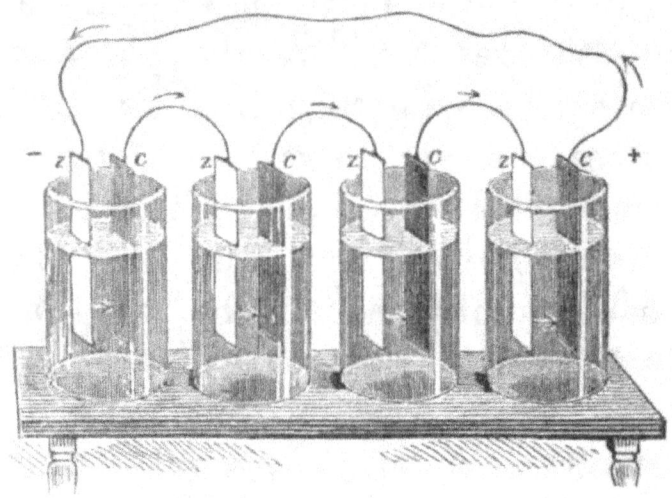

Fig. 13.—Battery of Simple Cells in series to show E.M.F. is proportional to number of cells.

Then if z is connected with the Earth, its potential will be zero, or will = 0. Let the *difference* of potential between the zinc and copper of this first cell be represented by the number 10; then the potential of the zinc of the *second* cell will also be 10 (for it is joined to c by a conductor, and must therefore have the same potential as c). But in this second cell also the *difference* of the charge upon the two metal plates is 10, and the potential of the copper plate in this cell will therefore be 20. But 20 is twice 10, so that the *two* cells will have double the E.M.F. of *one* cell. In the same way, with *three* cells we have *thrice*, and with *ten* cells *ten* times the electro-motive force of one cell. When cells are joined together in this manner they are said to be coupled in series; it is also called an *intensity* arrangement. The unit by which we measure electro-motive force is called the *volt* (after Professor Volta).

40. Advantage of Coupling Cells in Series.—A Daniell's cell, in good working order, has an electro-motive force of a little more than one volt; therefore twenty Daniell's cells, coupled in series, the copper of each cell being joined to the zinc of the next, will yield a force of over twenty volts. Grove's cell has an E.M.F. of a little less than two volts; so that the E.M.F. of twenty Grove's, coupled in series, would be rather under forty volts. The advantage of joining the cells of a battery *in series*, as described above, is that the additional electro-motive force thus obtained enables the current to overcome a greater external resistance. Thus, if we desire to send a fairly strong current through a great length of fine wire, or to produce an electric light (see Fig. 14), it will be desirable to arrange the cells in series, for in these cases the resistance to be overcome, outside the battery, will be considerable.

41. Reduction of Internal Resistance of a Voltaic Battery.—The plates, and more especially the liquids, of a voltaic battery offer a considerable *internal* resistance to the passage of the current. This internal resistance may be diminished (1) by bringing the plates nearer together, for then the current has to travel through less of the liquid; (2) by increasing the *size* of the plates, for there are then more paths by which the current can pass from one plate to another. Instead of actually increasing the

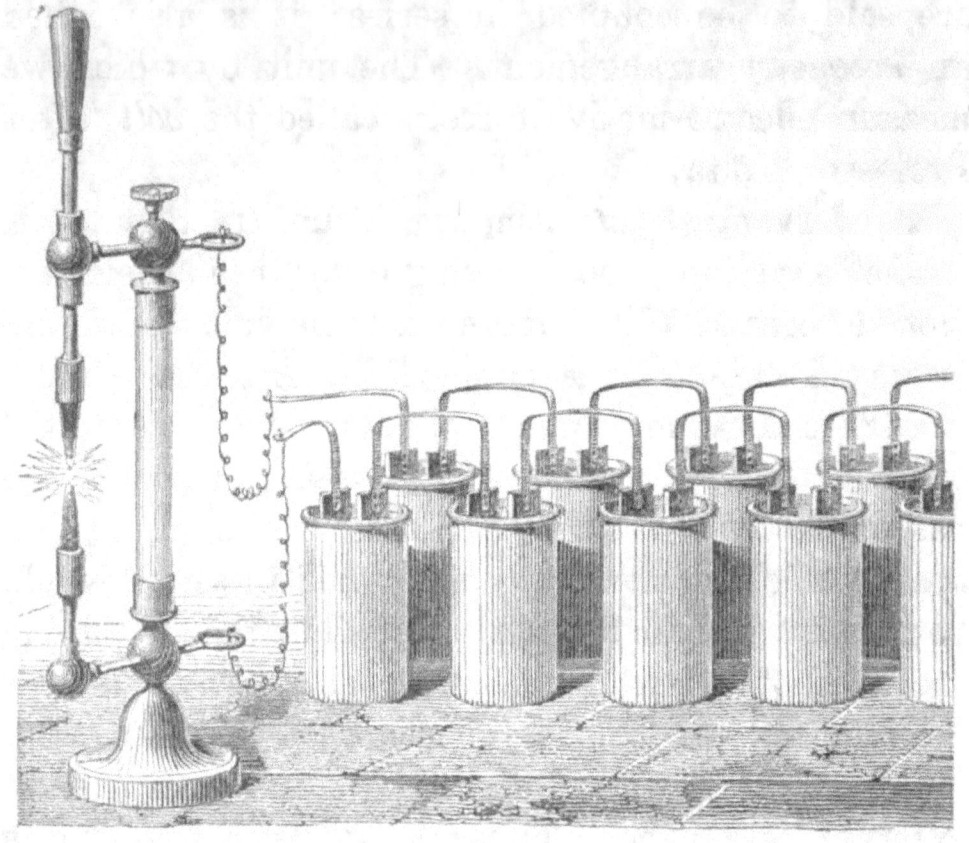

FIG. 14.—Battery of Cells in series to produce electric light.

size of the plates, it will have precisely the same effect if we join together the similar plates of different cells. Thus, if we have six cells, we may join the six zincs together, and also the six coppers, and the current will then be the same as that furnished by one large cell having a zinc plate equal in size to the six zincs, and a copper plate equal to the six coppers put together. When cells are joined in this way, they are said to be arranged in "compound circuit," or for *quantity* (see Fig. 15).

42. **Advantage of Coupling Cells for "Quantity."**—The E.M.F. of such a large cell would be only one-

sixth of that to be obtained by coupling the six cells "in series;" but then its internal resistance would only be one-sixth of that of the six cells. If the *external* resistance to be overcome by the current is but *small*, as in the case of heating a short length of platinum wire, then it will be advantageous to couple up the cells for "*quantity*." Generally speaking, the best effect is produced when the cells of a battery are so arranged that the total *internal* resistance is as nearly equal as possible to the total *external* resistance. Thus the manner in which the cells of a battery are best joined together depends entirely on the work which we desire the current to perform.

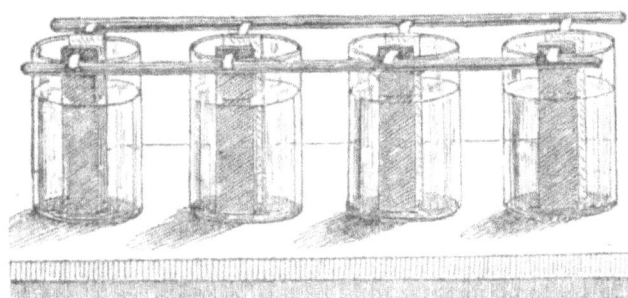

Fig. 15.—Battery of Simple Cells coupled for quantity.

VIII.—CHEMICAL ACTION OUTSIDE THE BATTERY.

43. Decomposition of Water by the Electric Current—44. How to effect the Decomposition of Water by Electricity: the Voltameter—45. Use of the Voltameter—46. Current of Polarization—47. Principles of Electrolysis—48. Electrolysis of Copper Sulphate—49. Electrotyping—50. Electro-plating.

43. Decomposition of Water by the Electric Current.—We have learned that chemical action takes place within each cell of a battery, and that by this means a current of electricity is maintained, chemical force being converted into electrical force. We have now to show that it is possible for chemical action to take place *outside* the battery when the electric current is caused to pass through certain liquids. Perhaps the simplest example is the decomposition of water by the voltaic current, an experiment first performed by Nicholson and Carlisle in the year 1800.

44. How to effect the Decomposition of Water by Electricity: the Voltameter.—The instrument generally employed to demonstrate the decomposition of water by electricity is known as a voltameter (Fig. 16). It consists of two strips of platinum placed in a small glass basin containing water, and connected by wires (which pass through the glass) with a

CHEMICAL ACTION OUTSIDE THE BATTERY. 47

voltaic battery, the cells of which are connected "in series." A few drops of sulphuric acid should be added, in order to improve the conducting power of the water. When the poles of a battery of two or three Grove's cells are connected with the binding-screws of the voltameter, the current is compelled to flow through the water in order to pass from the one strip of platinum to the other.

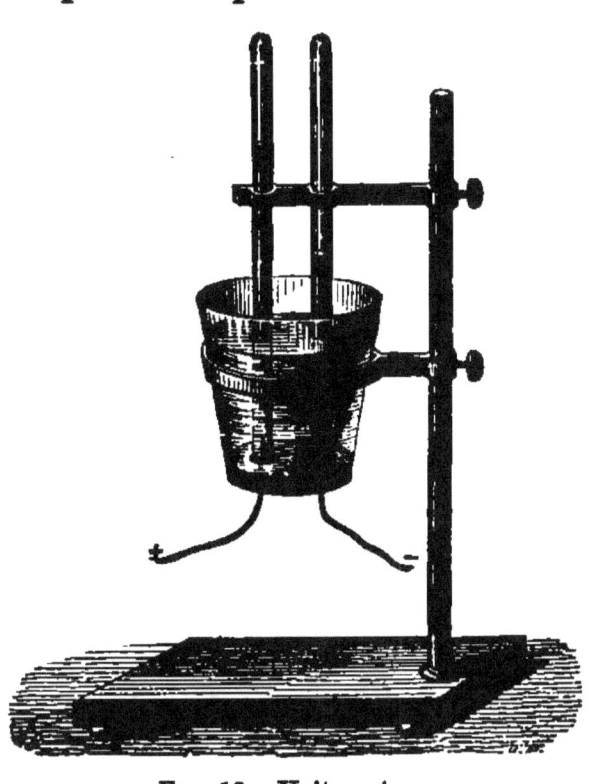

FIG. 16.—Voltameter.

The platinum connected with the positive pole of the battery becomes electrified +ly, while the other strip of platinum, connected with the negative pole, is charged −ly. Now the two elements which form water are charged, the hydrogen +ly, and the oxygen −ly. Thus the molecules of water which *lie between* the poles of the battery (as the strips of platinum may now be called) are drawn in opposite directions, the + H to the − pole, and the − O to the + pole. This electrical attraction is so strong that the atoms of hydrogen are torn away from the atoms of oxygen, and the water is decomposed. But there is no *chemical* affinity between either of these gases and the platinum, so that they rise in streams of bubbles from the surfaces of the metal strips, and may be collected in test-tubes previously filled with water and inverted, one tube over each strip of platinum. The tube up which the hydrogen gas rises will fill *twice as quickly* as the other tube, proving that there is twice as much hydrogen as oxygen in water. The decomposition of water may be represented by the following equation:—

$$H_2O = H_2 + O$$

Water *consists of* hydrogen (2 volumes) *and* oxygen (1 volume).

As each tube becomes filled with gas, it should be closed at one end with the finger, removed from the basin, and the gas tested by applying a light to it. The hydrogen will burn with a bluish flame; the oxygen will not burn, but it will relight a glowing match.

45. Use of the Voltameter.—The amount of water decomposed in any given time is proportional to the STRENGTH of the current, of which it therefore affords a measure. The glass tubes should be *graduated* (that is, marked upon the sides into cubic inches); then if the current from one battery produces a total of three cubic inches of the two gases in ten minutes, while that from a second battery gives six cubic inches in the same time, this proves that the second current is twice as strong as the first. For this reason such an instrument is called a voltameter (*volta*, a voltaic current; and *metron*, a measure), because it enables us to *measure* the strength of any voltaic current.

46. Current of Polarization.—After a battery current

has been flowing for some time through a voltameter, each of the strips of platinum becomes *polarized*, the strip where the current enters the liquid becoming coated with oxygen, while the strip facing it, where the current leaves the liquid, is covered with a film of hydrogen. If the wires leading from the voltameter to the battery be now disconnected or removed from the battery and attached to a galvanometer, a current of electricity will be found to flow in the opposite direction to the original battery current—namely, from the hydrogen film (which forms the + plate) to the oxygen film (the − plate), and thence through the wires to the hydrogen again. While this action continues, the voltameter may be regarded as a simple cell having plates composed of two gases immersed in an acid liquid, for the strips of platinum simply act as supports to the gases upon their surfaces. A *gas-battery* has in fact been constructed by Grove which acts in exactly the same way. If a number of water voltameters be included in the same circuit, and all become polarized by the current from a battery of many cells, they will yield a strong reverse current, which may be called a current of polarization.

47. Principles of Electrolysis.—The decomposition of any substance by means of an electric current is known as *electrolysis*, or electric analysis, as distinguished from chemical analysis, which effects the same object by different means. It was by electrolysis that water was first shown to be a compound. In the same way, in 1807, Sir Humphry Davy discovered the metals sodium and potassium by sending a powerful electric current through the substances known as soda and potash. These substances were previously thought to be *elements*, but Davy proved them to be compounds of a metal with oxygen. Thus—

$$Na_2O = Na_2 + O$$
Soda *is composed of* sodium *and* oxygen.

And in the same way—

$$K_2O = K_2 + O$$
Potash *is composed of* potassium *and* oxygen.

The word *electrolysis* is derived from the two Greek words *elektron*, electricity, and *lysis*, a loosening, and means the separation from each other, by the force of electricity, of the atoms which are combined to form the molecule of any compound. To effect the electrolysis of any substance, it is necessary (1) that the substance shall *not* be an element, (2) that it shall be in the *liquid* state, (3) it must be a conductor, and (4) a sufficiently strong current of electricity must be passed through it.

48. Electrolysis of Copper Sulphate.—When crystals of copper sulphate (blue vitriol) are dissolved in water we obtain a blue solution. Let the basin of a voltameter be now filled with this solution, and let a steady current from two or three Daniell's cells be passed through the liquid (see Fig. 17). The following decomposition takes place:—

$$CuSO_4 = Cu + SO_4$$
Copper sulphate *is decomposed into* copper *and* sulphion.

The copper is deposited upon the negative pole or

electrode, so that the platinum strip forming that pole soon becomes covered with a thin layer of metallic copper. But the current decomposes the *water* as well as the copper sulphate. The hydrogen from the water unites with the SO_4 (to which the name of "sulphion" has been applied) to form sulphuric acid (H_2SO_4), while bubbles of oxygen may be seen to rise from the positive pole. Thus the

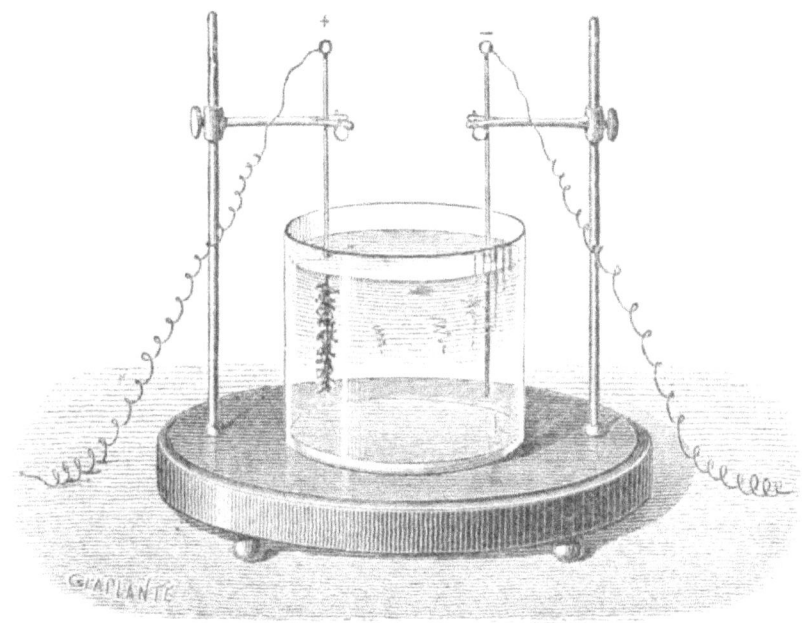

Fig. 17.—Decomposition of a salt by an electric current.

solution gradually becomes *acid*, and its blue colour disappears as more and more of the copper sulphate is decomposed. But if the positive pole be made of a plate of copper, then the solution does not lose in strength; for the sulphuric acid formed acts upon the suspended copper electrode, dissolving it to form more sulphate of copper. In this case it is plain that the negative pole will grow heavier (from the deposit of copper upon it), while the copper plate forming the positive pole will become lighter and lighter as it is dissolved by the newly-formed acid.

49. Electrotyping.—In the process of electrolysis it does not much matter of what material the negative pole is formed so long as it is a conductor. The metallic deposit upon it will continue to grow in thickness as long as the current flows through the solution. Advantage is taken of this to obtain *copies* of coins, ornaments, etc. First a *mould* of one face of the coin is taken in plaster-of-Paris. This

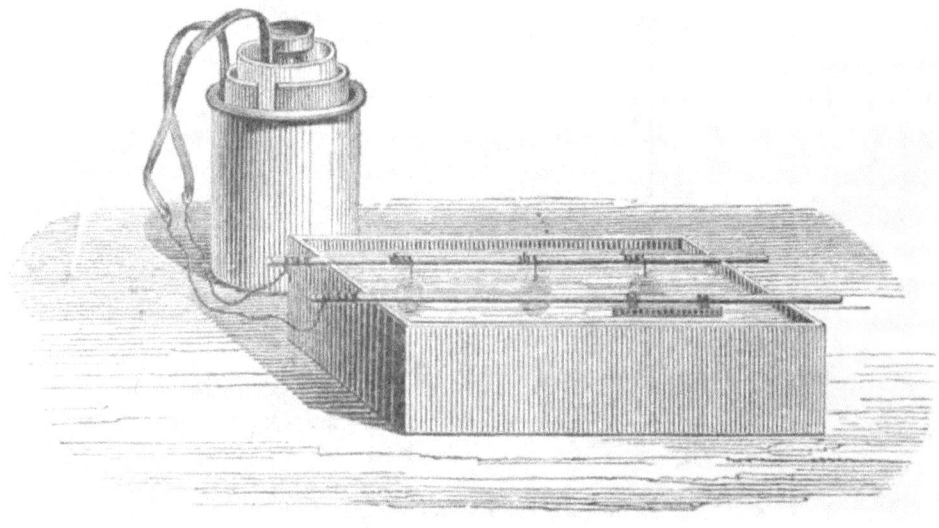

FIG. 18.—Electrotyping.

mould is soaked in melted paraffin, and the surface which is to receive the deposit is brushed over with black-lead to render it a conductor. The back and sides of the mould are varnished to *prevent* any deposit of metal upon them. The wire from the negative pole of the battery is then wrapped round the edge of the mould so as to be in contact with the black-lead. The mould is then dipped into a solution of copper sulphate contained in an earthenware vessel, and a steady current is allowed to flow for several hours (see Fig. 18). During this time the mould will receive a thick coating of copper, which can easily be detached from it, and which will be a perfect copy of the original object. The operation is called *electrotyping*, or the *electrotype process*; it was introduced in the year 1839.

50. Electro-plating.—In the year 1840, the Messrs. Elkington of Birmingham began to manufacture articles of German silver, to which they gave all the appearance of real silver by coating them with a

Fig. 19.—Electro-plating.

thin layer of that metal by means of a current of electricity. The objects—spoons, forks, etc.—are made extremely clean, and are then hung in a vessel containing a solution of the cyanides of silver and potassium. The wire from the copper plate (− pole) of a large Daniell's cell is then connected with the objects to be plated; while a plate of silver is fastened to the wire from the zinc (+ pole) and allowed to dip into the same solution opposite to the spoons, etc., which may be hung on a metal rod (Fig. 19). The time required for the completion of the process will depend on the thickness of coating desired, the strength of the current, and other considerations. The electro-deposition of silver is known as *electro-plating*. In a very similar way articles of silver, copper, brass, etc., may receive a coating of *gold*, the operation being then known as *electro-gilding*.

IX.—HISTORY OF CURRENT ELECTRICITY.

51. Muscular Motion produced by the Return Shock—52. Muscular Motion produced by the Action of Metals—53. Volta's Explanation of Galvani's Experiment—54. Construction of the Voltaic Pile—55. Action of the Voltaic Pile—56. The "Couronne des Tasses"—57. Sulzer's Experiment—58. Production of a Voltaic Current by Simple Means.

51. Muscular Motion produced by the Return Shock.—The return shock is a phenomenon well known to students of frictional electricity. When the prime conductor of a large plate machine is highly charged with, say, positive electricity, it acts inductively on neighbouring objects, repelling their positive and charging them with negative electricity. But when the prime conductor is discharged, the repelled positive rushes back, and the surrounding bodies resume their neutral state. To this recombination of the two electrical fluids, which have been separated by induction, the name of the "return shock" is given.

In the year 1780, Galvani, professor of anatomy at Bologna in Italy, noticed that the return shock could produce muscular contractions of the limbs of animals. A recently killed and skinned frog, lying upon a table near an electrical machine, was observed to move convulsively every time a spark was drawn from the prime conductor of the machine. This fact greatly interested Galvani, and he experimented in many ways upon the bodies of animals, believing and endeavouring to prove that all the motions of the muscles of animals, and the actions of their nerves, are due to a kind of fluid resembling electricity, to which he gave the name of the "vital fluid."

52. Muscular Motion produced by the Action of Metals.—Galvani conducted his experiments for six years with but little success. But in 1786 he was rewarded by a new and very important discovery. Endeavouring to show that the return shock sometimes produced by the action of electrified clouds during a thunderstorm would have the same effect upon a frog's muscles as that produced by a cylinder machine, he suspended the legs of a frog (by means of a copper wire passed through the spinal cord) to the iron railings outside his house. By the force of the wind the lower part of the frog was occasionally made to touch the iron uprights, and whenever this happened, although there were no thunder-clouds about, a violent movement of the legs took place. Galvani explained this movement by saying that the nerve (through which the copper wire passed) was charged with one kind of electricity, and the muscle of the leg with another, and that the two metals—the copper and the iron—allowed the two electric fluids to join one another, in doing which they produced muscular motion. This theory was believed in for some time, but we now know that it is not true. Galvani's experiment may easily be repeated with the apparatus shown in Fig. 20.

53. Volta's Explanation of Galvani's Experiment.—To Professor Volta of Pavia belongs the honour of offering a correct explanation of the motion observed when a muscle of an animal is connected

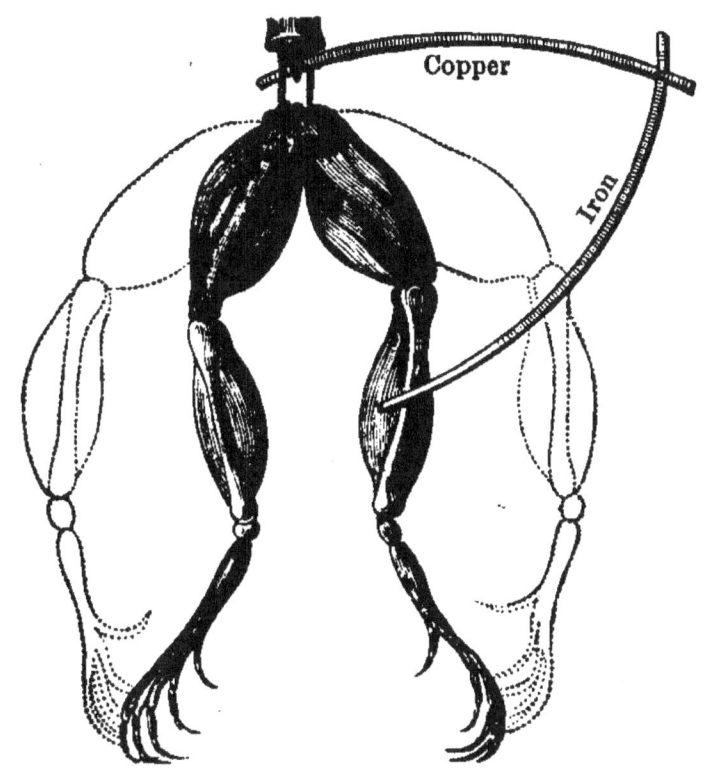

Fig. 20.—Nerve and muscle of frog touched by compound metal bar.

with a nerve by means of two distinct metals. Volta showed that when any two metals, such as copper and iron, touch one another, the copper becomes charged with negative and the iron with positive electricity. The moist body of the frog—the nerve and the muscle—serves as a conductor to allow these two electricities to unite. But, in passing, the electric current stimulates the nerve, and the nerve then causes the muscle to contract. This is Volta's famous *contact theory* of electricity, to which we have already alluded. Volta's explanation is the exact opposite of Galvani's. The latter thought the electricity was produced in the frog, but Volta showed that it had its origin at the place where the metals touch one another. Galvani thought that the metals merely conducted the electric current, but Volta proved that it was the moist tissues of the frog by which this was accomplished.

54. Construction of the Voltaic Pile.—Continuing his experiments, Volta found that when the body of the frog was replaced by some substance saturated with an acid, or with some saline liquid (such as brine), even stronger currents of electricity passed from the one metal to the other. He also found it convenient to connect the two metals by a wire, as it was then more easy to study the properties of the electric current as it flowed along the wire. To prove this, let a piece of flannel, or two or three thicknesses of blotting-paper, be saturated with acid water; place pieces of zinc and of copper upon the flannel, and connect them by a wire; then a feeble voltaic current will be found to traverse the wire.

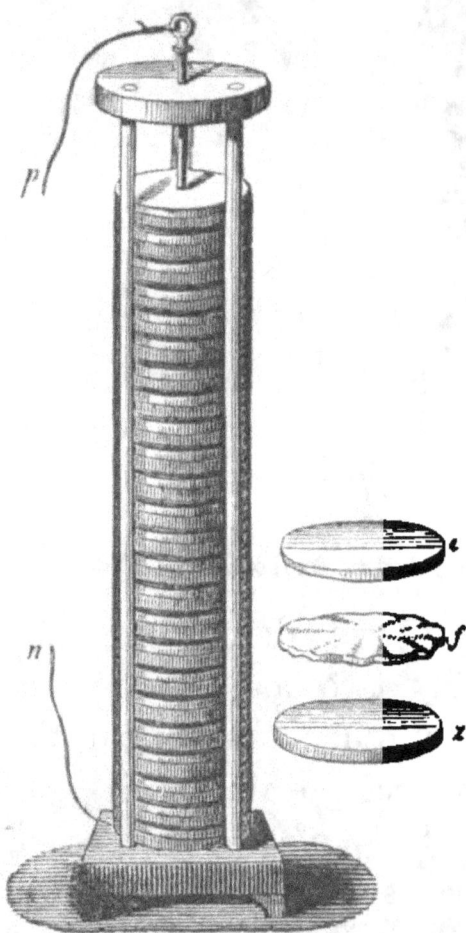

FIG. 21.—Voltaic Pile. z, Zinc disc; f, flannel; c, copper disc; p and n are wires connected with the + and the − poles of the pile respectively.

To increase the strength of the current, Volta *piled up* his metal discs, placing first a disc of zinc, upon which he laid a disc of copper, and then a rather larger circular piece of flannel; then came other pieces of zinc, copper, and flannel, all arranged in the same order (Fig. 21). The entire pile was kept upright by means of glass rods fastened to a stand below and to a wooden cap above. It is now usual to solder together the discs of copper and zinc, as this secures perfect contact between the two metals, and enables thinner discs of each to be employed.

55. Action of the Voltaic Pile.—When a voltaic pile of about sixty pairs of metal discs is employed, an electric current of considerable intensity traverses the wire by which the highest zinc is connected with the lowest copper. On consideration it will be seen that the top zinc disc is merely a metallic extension of the copper upon which it rests, and the copper plate at the bottom of the pile is in the same way to be considered as part of the zinc plate which rests on it. Remembering this, the path of the current, *up* the pile and out of the highest copper disc, through the wire to the zinc pole at the bottom of the pile, will be understood; for we have learned that in all batteries (and the voltaic pile is really a battery) composed of zinc and copper plates, the positive current *issues from* the *copper* pole. When the two ends of a freshly set up voltaic pile, or the wires proceeding from them, are simultaneously touched with wetted fingers, a distinct shock will be felt, for the difference of potential is considerable. An experimental voltaic pile may be built up of coins, using pennies and half-crowns for example, with discs of blotting-paper dipped in brine placed between them.

It was in the year 1800 that Volta invented this "voltaic pile," and the discovery marks an era in the progress of the science of electricity. For the first time the power of producing a *continuous* current was placed in the hands of electricians, and the results have indeed been great and wonderful.

56. **The "Couronne des Tasses" (or Crown of Cups).**—The current from a voltaic pile decreases very rapidly in strength. One reason is that the weight of the metal discs squeezes out the acid liquid from the flannel, and when the flannel thus dries up there is nothing to sustain the current. A partial remedy for this is to lay the pile on its side; but Volta soon found that it was better to do away with the flannel altogether, and to place the strips of metal in glass vessels which could hold a considerable quantity of the exciting liquid. Taking several small glass cups or basins, about three parts full of acid water, he placed in each of them a strip of copper and a strip of zinc (Fig. 22), keeping the metals an inch or so apart. He then joined the copper strip of one cup to the zinc of the next by a short piece of wire, the copper of that cup to the zinc of the next, and so on. When the zinc of the first cup was finally connected with the copper in the last cup, the wire by which they were joined was traversed by an electric current which was far more powerful and more lasting than the current yielded by the voltaic pile. Since the glass vessels were usually arranged in a circle, Volta called them his "couronne des tasses," which

means in English a "crown of cups." It is plain that each cup, with its acid liquid and pair of dissimilar metals, is neither more nor less than a SIMPLE CELL.

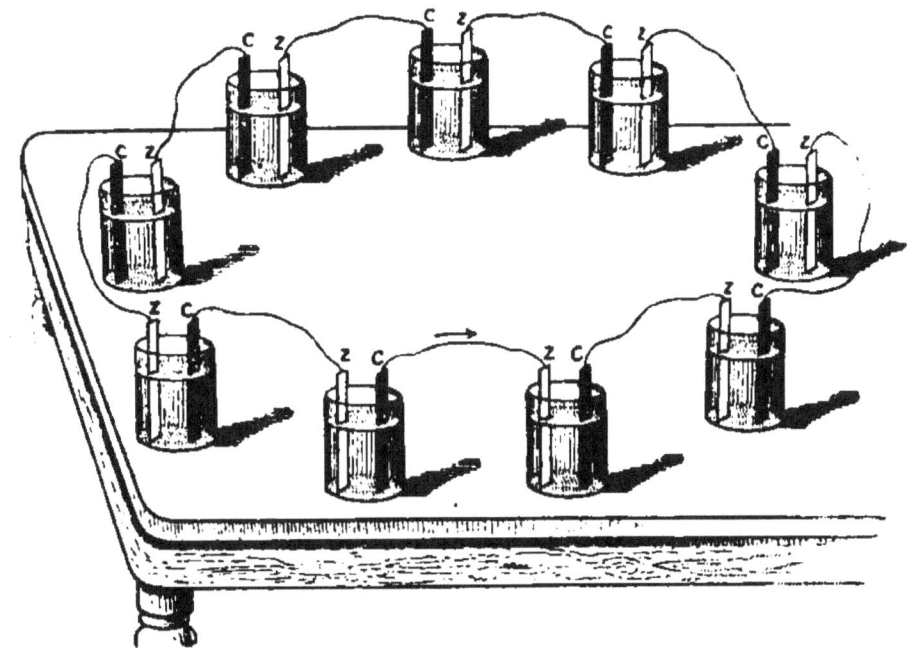

Fig. 22.—Couronne des Tasses.

57. Sulzer's Experiment.—In 1752, a German electrician, named Sulzer, noticed that if one metal, as a silver coin, be placed on the top of the tongue, while some other metal—a bit of zinc or a steel pen for example—is placed underneath the tongue, a curious and unpleasant *taste* will be perceived when the ends of the two pieces of metal are made to touch one another. If we connect each piece of metal by means of a wire with a delicate galvanometer (Fig. 28), it is easy to show that a voltaic current travels along the wire. The experiment is, in fact, very similar to that first noticed by Galvani, the moist tongue representing the body of the frog. The precise cause of the *galvanic taste* is not yet known, but that it is due to a current of electricity is shown by the fact that the same taste is produced when the two wires from a voltaic cell are placed in contact with the tongue.

58. Production of a Voltaic Current by Simple Means.—To produce a voltaic current, it is only necessary to immerse two dissimilar metals in a liquid which is capable of acting more upon one of the metals than upon the other, and then to connect the two metals by some substance capable of conducting electricity. To detect feeble currents, it will be needful to place an astatic galvanometer (Fig. 28) in the circuit. The following examples of currents produced by simple means may be given:—

(1.) Place a pin (having a surface of metallic tin) and a needle (iron) in an orange, and connect each by a fine wire with a binding-screw of a galvanometer. The acid juice will corrode the tin, and a feeble current will flow from the tin to the iron (through the orange), and thence along the wire through the galvanometer to the tin again.

(2.) In the same way a weak voltaic current is produced when a silver fork and a steel knife are plunged into a juicy beefsteak which has been well sprinkled with salt.

X.—MAGNETIC EFFECTS OF THE VOLTAIC CURRENT.

59. Connection between Magnetism and Electricity—60. Oersted's Discovery—61. Ampère's Rule—62. Simple Galvanometer—63. Multiplying Galvanometer—64. Astatic Galvanometer—65. Insulation of Wires—66. Magnetic Effect of Current on Iron Filings.

59. Connection between Magnetism and Electricity.—Experimenters long suspected, before they were able to prove, that there was some connection between the force of magnetism and the force of electricity. It was known, for example, that in thunderstorms steel objects, such as knives, had been magnetized by the passage through them of lightning. It was found that by sending the discharge of a Leyden battery through a strip of tinfoil, a sewing-needle placed across the strip was also magnetized.

60. Oersted's Discovery.—It was the good fortune of a Dane, named Oersted, to discover, in the year 1819, the remarkable effect produced upon a balanced magnet by the passage near it of an electric current.

(1.) When a wire through which a current is flowing from north to south is placed parallel to and *above* a magnetic needle, the north pole of the needle moves towards the *east* (Fig. 23).

64 MAGNETIC EFFECTS OF THE VOLTAIC CURRENT.

(2.) If the wire is placed so that the current flows from south to north above the needle, then the needle's north pole moves to the *west*.

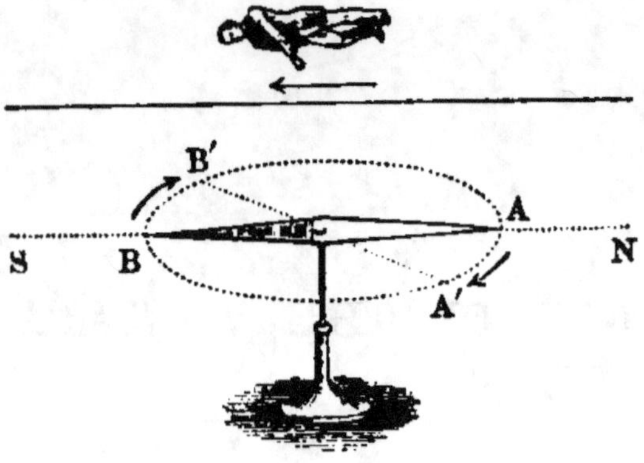

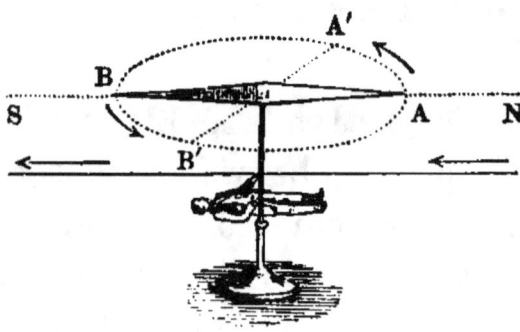

Fig. 23.—Action of current on magnet.

Fig. 24.—Action of current on magnet.

Placing the wire *underneath* the needle, exactly the opposite movements are observed.

(3.) When the current flows from north to south *below* the needle, the N.-seeking pole of the magnet is deflected to the *west* (Fig. 24).

(4.) When the current flows from south to north below the needle, it causes the needle's N.-seeking pole to move towards the *east*.

61. Ampère's Rule.—As an easy way of remembering the direction in which the north pole of a magnet would be deflected by a voltaic current, the French electrician Ampère suggested the following amusing but useful rule: *Imagine a little figure of a man to lie within the wire, always facing towards the needle and swimming with the current; then the north pole of the needle will always be deflected to his* LEFT *hand* (see Figs. 23, 24).

62. Simple Galvanometer.—The word *galvanometer* is derived from *galvano*, a current of galvanic

MAGNETIC EFFECTS OF THE VOLTAIC CURRENT. 65

electricity, and *metron*, a measure. The instrument known by this name consists of a magnetic needle carefully balanced on a pivot, having underneath it a circular card marked with degrees, while a wire is placed above and parallel to the needle, at a distance from it of about one inch. If we connect the ends of this wire with the poles of a voltaic cell, the needle will be deflected; and the stronger the current the greater will be the deflection. By noticing the number of degrees over which the N.-seeking pole is repelled, we can compare the strength of any one current with that of any other.

63. Multiplying Galvanometer.—Suppose that the wire bearing the current is bent round so that the current flows *below* the magnet *as well as above it* (Fig. 25), then it is easy to see that the current in each part of the wire will urge the N.-seeking pole in the

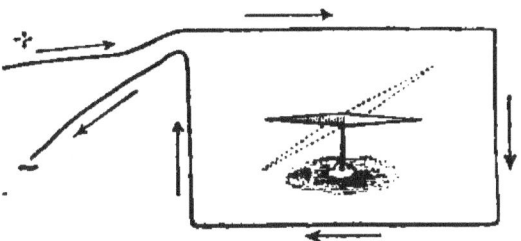

FIG. 25.—Current flowing above and below needle.

same direction; for if the current in the wire above the needle be flowing from south to north, then (by Ampère's rule) a man swimming with the current and facing the needle would have his left hand to the *west*. And the same man, swimming on with the current, would, when in the wire *below* the needle, have to lie on his back to face the magnet, and in that case also the west would lie to his left hand. This can be made evident by cutting out a little paper figure of a man and sliding it along the wire in the direction of the current. If the wire is supported by a wooden framework (Fig. 26), it may be coiled ten, twenty, or a hundred times round the pointed needle, and the greater the number of turns the more powerful will be the effect of the current upon the magnet. Since the effect of the current is *multiplied* by increasing the number of turns of the wire, this instrument is known as the multiplying galvanometer. But the increase in the number of turns requires an increase in the total *length* of the wire, and this increased length causes increased *resistance* to the passage of the electricity. Thus, if the electro-motive force by which the current is urged is but small, it will be better to use a short, rather thick, wire, making thirty or forty turns only. For comparing and measuring the strength of currents produced by chemical action, as in the different kinds of voltaic cells, from five hundred to one thousand turns may be used with advantage.

Fig. 26.—Multiplying Galvanometer.

64. Astatic Galvanometer.—An astatic combination consists of two magnetic needles placed with unlike poles one above the other, and fastened together by a piece of copper wire (Fig. 27). Such a pair, if the needles be of equal magnetic strength, will remain at rest in any position, the

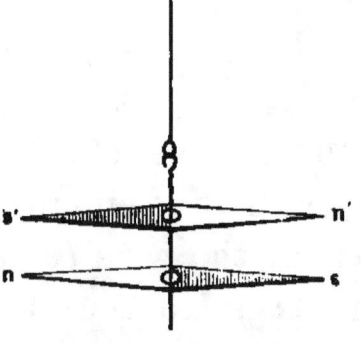

Fig. 27.—Astatic combination of two needles.

Earth's magnetism having no effect upon it. Now a single magnetic needle tends to remain in a north and south position, and the voltaic current has to overcome the force of the Earth's magnetism before it can deflect the needle. With the astatic pair no such force, or but a fraction of it, has to be overcome. Hence, to detect and measure weak currents, an astatic galvanometer is to be preferred. The needles are usually suspended by a fine thread, so that one hangs within and the other above the coil of wire, and the instrument is placed under a glass shade (Fig. 28) to screen the needles from currents of air.

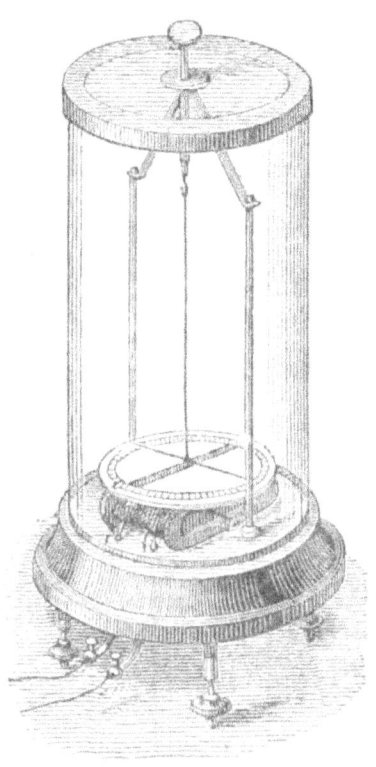

Fig. 28.—Astatic Galvanometer.

65. Insulation of Wires.—To prevent the escape of electricity from a metal wire, it is usual to surround the wire with some substance which is a non-conductor—that is, one along which the electricity cannot pass. For many purposes a covering of white cotton is sufficient; but silk is a far better insulator than cotton, so that copper wire is frequently covered with green silk. For thick wires a coating of gutta-percha is preferred. The wire of a multiplying galvanometer *must* be insulated, so that the electricity cannot pass from one layer to another. If this were not done, any number of turns of bare wire, resting one upon the other, would only produce the effect of a single turn.

66. Magnetic Effect of Current on Iron Filings.—When a steel knitting-needle is magnetized and then rolled in iron filings, the filings are attracted by the magnetic force possessed by the steel, and cluster round it. It is not possible to magnetize a copper wire in the same way that we can magnetize a bar of steel—by rubbing it with a magnet; but let a powerful voltaic current be made to flow through the copper wire, and, *while the current is flowing*, the wire will be found to possess certain magnetic properties. To show the magnetic effect of the electric fluid a tolerably strong current is required, such as would be produced by, say, three Grove's cells. If the copper wire through which such a current is flowing be dipped into iron filings, many of the filings will adhere to the wire, just as if it were a magnet. When the circuit is broken, and the current ceases to flow, the filings drop off. That the wire conveying the current is surrounded by a *magnetic field* may be shown by passing a vertical wire through a card; iron filings strewn upon the card arrange themselves in circles round the wire (Fig. 29).

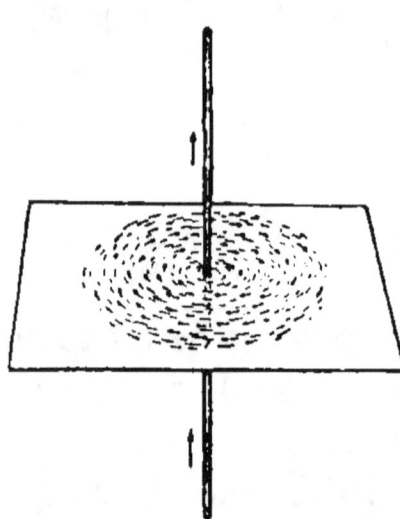

Fig. 29.—Wire passed through card: filings in curves.

XI.—ELECTRO-MAGNETS.

67. Solenoid—68. Floating Solenoid—69. Magnetization of Soft Iron—70. Construction of Electro-magnets—71. Uses of Electro-magnets.

67. Solenoid.—We have seen that a copper wire traversed by a voltaic current resembles a magnet in its power of attracting iron filings. If the wire is wound into a spiral coil, as by wrapping it round and round a ruler, and the two ends are then brought back inside the coil and out at its centre, we shall have a *solenoid*, which will act in *all respects* like a magnet when the wire (which should be well insulated by a covering of silk or of gutta-percha) is traversed by a voltaic current (Fig. 30).

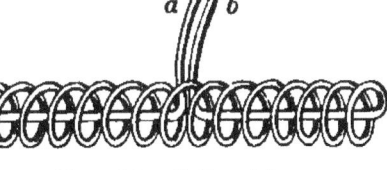

FIG. 30.—Solenoid.

If the coil is suspended so that it can turn freely in any direction, which may be done by bending the ends, *a b*, of the wire, and causing them to dip into little cups of mercury with which the poles of the battery are also connected, then the solenoid will set itself north and south, just like a magnet. It can attract or repel other magnets (according as unlike or like poles are brought near), and in the same way it can attract or repel another solenoid.

68. Floating Solenoid.—If the two ends of the wire of a solenoid be pushed through a bung, and a plate of copper fastened to one end and a plate of zinc to the other end of the wire, the instrument may be placed in a basin of acid water, and we shall have a *floating solenoid*, combining battery and coil in one, which will act just like a floating magnet (Fig. 31). Although the magnetic power of a solenoid is quite perceptible, yet it is very weak.

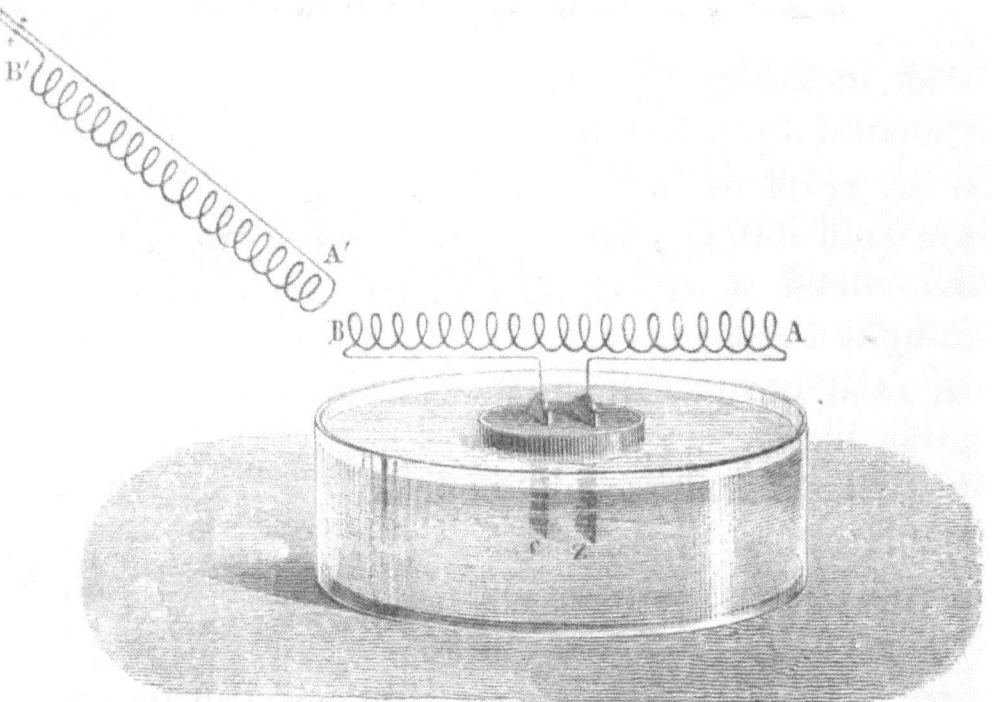

FIG. 31.—Floating Solenoid A B attracted (like a magnet) by the second Solenoid A' B'.

69. Magnetization of Soft Iron.—In the year 1820, Arago, a famous Frenchman, and Sir Humphry Davy, the great English chemist, discovered that when a voltaic current was made to travel, by means of a well-insulated wire, round and round a bar of soft iron, the soft iron became a temporary magnet. When the circuit was broken, and the

ELECTRO-MAGNETS. 71

current ceased to flow, the soft iron lost all its magnetic properties.

Let the soft iron nail, *n s* (Fig. 32), be placed below the wire, *a b*, along which a strong current is flowing from *a* to *b*; then the nail will be magnetized by the current, so that its north-seeking pole will be at *n*. In fact, we can again apply Ampère's rule to the discovery of the poles of electro-magnets: "*Imagine a little figure of a man to lie within the wire, swimming with the current and facing towards the iron bar; then the north pole of the bar will be to his* LEFT *hand.*"

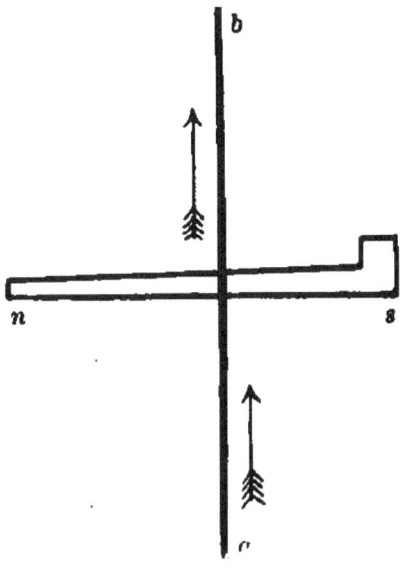

FIG. 32.—Magnetization of soft iron by electric current.

70. Construction of Electro-magnets.—The iron bars or cores most suitable for electro-magnets are made of very *soft* or pure iron, such as the kind known as Swedish charcoal iron. These bars should be carefully *annealed*—that is, they should be made red-hot in a furnace, and then allowed to cool *very slowly*, as by allowing them to cool down in and along with a heap of hot ashes.

Commencing at one end of the soft iron bar, the wire must be carefully coiled round and round, always in the same direction (Fig. 33). We may learn and remember the poles of an electro-magnet by noticing that round the *south* pole the current moves in the *same* direction as the hands of a watch, but flows in the *opposite* direction to watch-hands round the magnet's *north* pole.

The coils of wire must be insulated both from the iron bar and from each other, or the current will *short circuit*—that is, will take the nearest path directly along the bar, instead of passing round and round it, as we desire.

FIG. 33.—Horse-shoe electro-magnet.

The most useful form of electro-magnet is that in which the iron bar or core is bent into the form of a horse-shoe (Figs. 33 and 34), so that the two poles can *act together* in attracting the *keeper*, as the piece of soft iron by which the two poles are then connected is known. It is best to coil the wire thickly, coil upon coil, round the two "legs" of the core, leaving the middle part bare. In this way electro-magnets have been made which are able to sustain a weight of several tons, hung from the keeper.

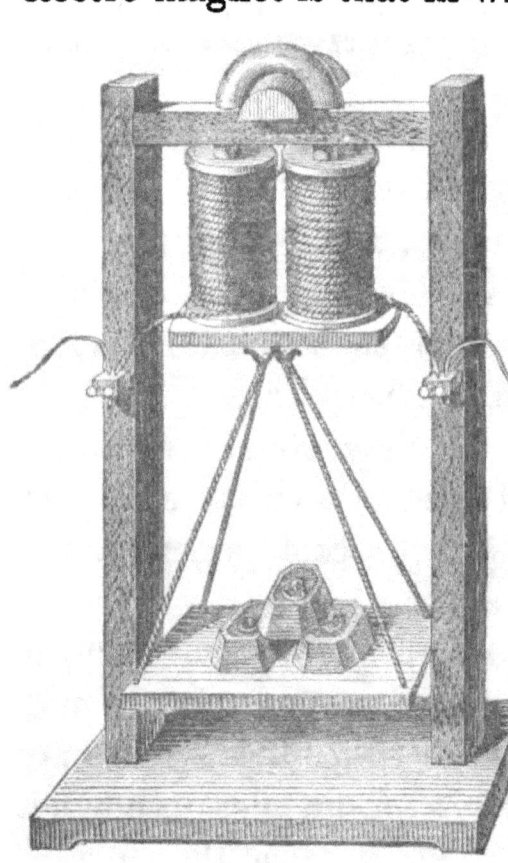

FIG. 34.—Lifting-power of an electro magnet.

71. Uses of Electro-magnets.—Electro-magnets usually far exceed in power ordinary steel magnets of the same weight.

(1.) They are much used for separating particles of iron from other substances with which they may be mixed. Brass-founders and iron-smelters employ them for this purpose.

(2.) One advantage of an electro-magnet is that it can be made to acquire or to lose its magnetic power at our will and from a distance. It may thus be made to sound a bell at almost any distance, by fixing in a wooden box the electro-magnet, the bell, and a piece of soft iron so placed that it will strike the bell when the keeper is attracted by the magnet (Fig. 35). A battery of two or three cells must be connected with the wire passing round the electro-magnet. Then when the circuit is completed the iron will be attracted, and the bell will sound. When the circuit is broken the iron will be released, and is made to move back to its first position by means of a small steel spring.

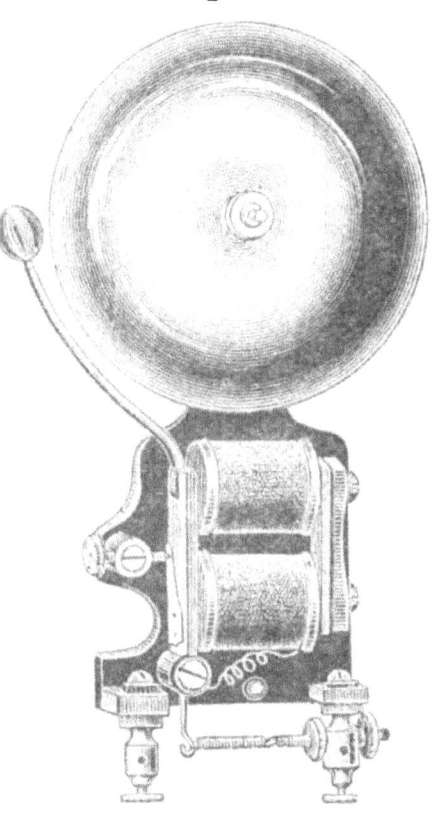

Fig. 35.—Electric bell.

(3.) Electro-magnets are used in the construction of the dynamos by which the electric light for our streets and public buildings is produced.

XII.—THE ELECTRIC TELEGRAPH.

72. History of the Electric Telegraph—73. Working of the Electric Telegraph—74. The Transmitter—75. The Receiver—76. Submarine Cables—77. Detection of Currents in Telegraph Wires.

72. History of the Electric Telegraph.—Our ordinary system of telegraphy depends upon Oersted's discovery of the deflection of a magnetic needle by an electric current. The distance of the needle from the voltaic battery (the source of the current) is of small importance so long as the two poles of the battery are connected with the two ends of the wire which passes round the needle; the battery may then be in London and the needle in York, still the north-seeking pole will turn westward when the current passes round it in one direction, and eastward when the current travels the opposite way along the wire. Nothing will then be more easy than to agree beforehand on a code of signals by which we can spell out words: one turn of the needle to the right may indicate the letter a, one to the left b, two turns or deflections to the right c, and so on. The sender, stationed, say, in London, can cause the current from his battery to travel in which direction he pleases, and as frequently as he pleases, along the wire; the needle in York will move accordingly, and the observer who is watching it will understand the message which is thus being sent. The electric telegraph has the advantage of outstripping all other means of communication. The speed at which the current travels is so great that the time it takes to pass, for example, from London to York is only a very small fraction of a second.

It is thought that Ampère, at Paris in 1821, first suggested the employment of a galvanometer in a circuit as a means of transmitting messages from one place to another. This idea was worked out by certain German electricians, and the needle telegraph was invented and introduced into England by Cooke and Wheatstone in 1837. Before this time the important discovery had been made that a *single* wire was sufficient to carry the current from the battery to the needle, for the surface of the Earth, or rather the soil, serves instead of, or can take the place of, the second or return wire.

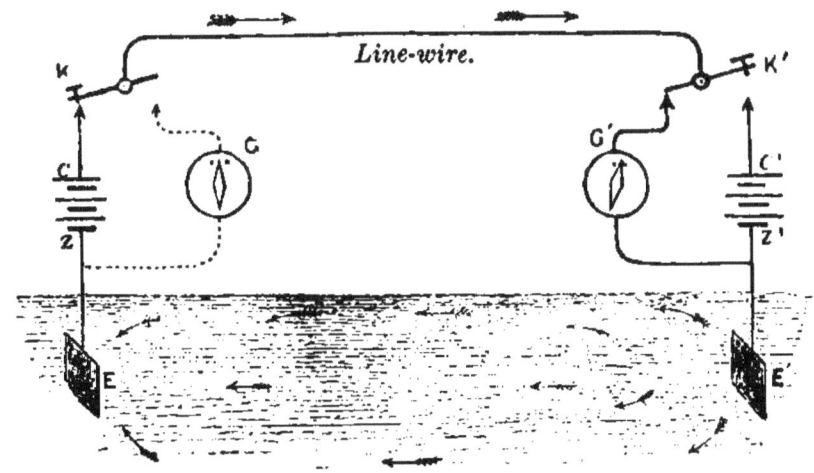

FIG. 36.—Electric telegraph, line-wire, etc.

73. Working of the Electric Telegraph.—In Fig. 36 we see a diagram of the arrangement of a telegraph line. The battery, C Z, has its zinc pole connected with a large metal plate, E, buried deeply in the ground, while the copper plate, C, can be connected with the line-wire by pressing down the key at K. The line-wire is usually made of galvanized iron, and is supported by stout posts to which insulators made of porcelain are screwed, and the wire is fastened to these insulators, so that it does not touch the posts or any other conducting substances. In this way the electricity is prevented from escaping from the wire, and is forced to travel on to the galvanometer, G', where it produces a deflection of the needle. The current then passes down to the second earth-plate shown at E', and returns through the ground, in the direction shown by the arrows, to the place from whence it started.

The diagram thus shows the instruments required to send messages from London to York. For messages in the opposite direction—from York to London—there would of course be required another battery, C' Z', at York, and another galvanometer, G, at London.

74. The Transmitter.—The instrument by which currents are sent from the battery is known as a signalling key, or transmitter, or "tapper." It consists of two brass keys or levers (Fig. 37), one of which is connected with the "air-line" (as the wire attached to the telegraph posts is called) and the other with the earth-plate. The same wooden base which supports these levers also carries two horizontal brass plates, the upper one connected with the zinc end (− pole) of the battery, and the other with the copper end (+ pole). When not in use the two levers are both in contact with the upper or negative pole. By pressing down the "line key" the circuit is completed, and the current passes through the air-line, round the distant galvanometer, and back by the earth. Releasing the line key, and pressing down the "earth key," the current travels in exactly the opposite direction — namely, through the earth, and back by the air-line.

FIG. 37.—The Transmitter, needle telegraph.

75. The Receiver.—The receiving instrument for ordinary telegraphic messages is simply an upright galvanometer (Fig. 38). A lightly-hung magnetic needle is surrounded by a coil of the line-wire, and is caused to move by currents travelling along that wire from a battery which may be hundreds or thousands of miles distant. The needle which is visible in front of the dial is merely an index, or pointer, fastened to and moving with the real magnetic needle, which stands within the coil behind the dial. For convenience, the receiver is usually

78   THE ELECTRIC TELEGRAPH.

placed upon the same base as the transmitter; but of course the tapper at one station works the receiver at another.

76. Submarine Cables.—Between countries which are separated from one another by water, as by a strait or an ocean, telegraphic communication is carried on through *cables*, which lie upon the bottom of the sea. In Figs. 39, 40, we see the exterior and a section of such a cable, showing four strands of pure copper wire in the centre, surrounded by an insulating coat of gutta-percha, outside which comes a protecting sheath composed of iron wires covered with hemp. The first submarine cable was laid between Dover and Calais in 1851; and the great task of connecting England and America by a cable 3,500 miles in length, resting on the soft white mud which forms the floor of the North Atlantic Ocean, was successfully accomplished in 1866.

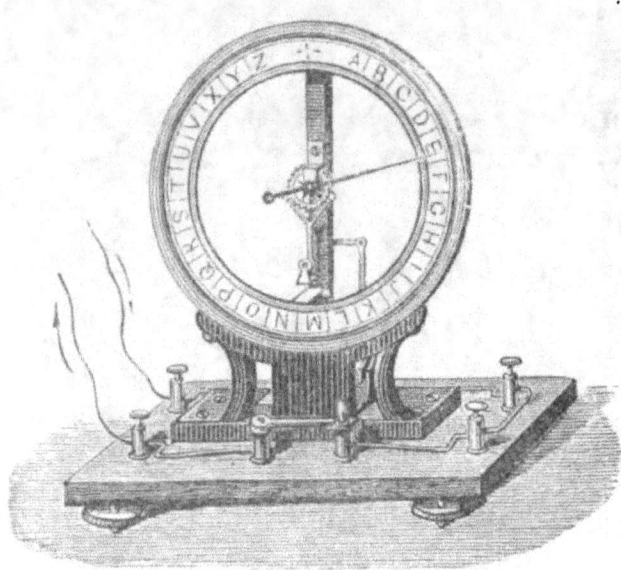

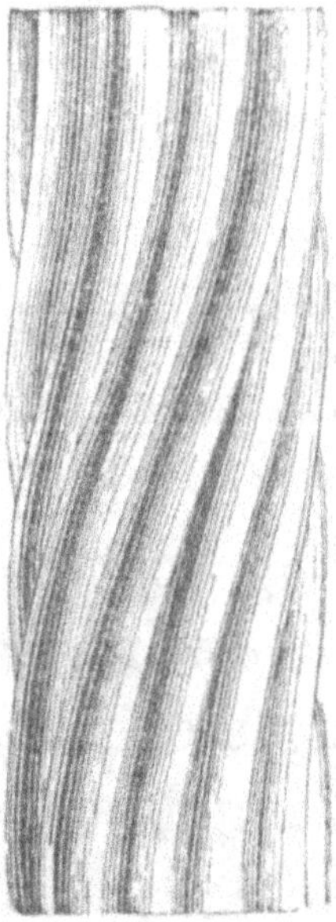

FIG. 38.—Receiver (alphabet form).

FIG. 39.—Submarine Cable.

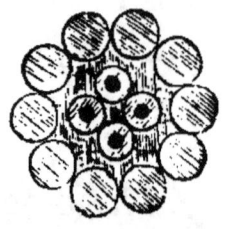

FIG. 40.—Section of Submarine Cable.

77. Detection of Currents in Telegraph Wires.—Provided with a well-balanced magnetic needle—a compass, for example—we should easily be able to tell whenever a telegraphic message was passing along a line-wire which ran north and south. We should have to climb up one of the posts and rest the compass upon the wire, when the passage of a current along the wire would be made evident by a deflection of the needle. Moreover, if the north pole of the needle moved westward, we should know that the current was flowing from north to south; if the north pole moved to the east, that would prove that the direction of the current in the wire beneath was from south to north.

If the line-wire runs east and west, or in any other direction than north and south, we might (1) use an *astatic* needle, which would remain at rest parallel to the wire when no current was flowing, or (2) turn the compass on its edge and present one *end* of the needle to the side of the wire. Then the north pole would move *upward* if the current ran from east to west, *downward* if it passed from west to east, Ampère's rule still holding good.

XIII.—HEATING EFFECTS OF THE CURRENT.

78. Influence of Temperature upon Resistance—79. Production of Heat in a Voltaic Circuit—80. Heat produced within the Battery—81. Heat produced outside the Battery—82. Relation of Temperature to Current—83. Relation of Temperature to Resistance—84. Experiments with Platinum and Silver—85. Blasting by Electricity.

78. Influence of Temperature upon Resistance.—When a piece of metal—as a copper wire—is heated, it does not conduct electricity so well as when it is cold. The effect of heat upon a metallic conductor is invariably to *increase* its resistance. This may be proved by heating with a spirit-lamp or a Bunsen burner a portion of the wire by which a voltaic battery is connected with a galvanometer. The increased resistance of the hot wire diminishes the flow of electricity, and this is shown by the *smaller* angle which the galvanometer needle then makes with its coil of wire.

79. Production of Heat in a Voltaic Circuit.—When a substance combines with oxygen, chemical energy is produced. Now, in a voltaic cell the oxidation of the zinc is a continual source of such energy. The energy so produced is distributed over the entire circuit in accordance with the resistance offered by each part of the circuit. Where the resistance is great, the heat developed by the energy is great; and where there is little or no resistance to the passage of the current, there the energy of the current develops little or no heat.

80. Heat produced within the Battery.—If the two poles of a battery be connected by a short, stout wire made of some good conducting material, then the heat is developed almost entirely *within* the battery, the metal plates and the liquids between them quickly becoming warm.

81. Heat produced outside the Battery.—The battery is, as it were, the hearth on which the zinc is burned; but if the two poles be connected by a long and thin wire (which will offer considerable resistance to the passage of the current), then this wire will become heated, while *less* heat will be produced *within* the battery. The *total* amount of heat produced will, however, be the same.

82. Relation of Temperature to Current.—All conductors are heated by the passage through them of an electric current. But if the conductor be very thick, and of good conducting material—as silver or copper—the heat produced may be hardly perceptible. The heat produced is proportional, in the first place, to the *square* of the strength of the current. Double the current, and we get *four* times the amount of heat; make the current *three* times stronger, and it will produce *nine* times as much heat, and so on.

83. Relation of Temperature to Resistance.—In the second place, the heat produced is exactly proportional to the electrical *resistance* of the wire or other conductor through which the current passes. If the substance of which the wire is made is a good conductor of electricity, then it will be *less* heated by the passage of a current than a similar wire made of a bad conducting substance would be.

84. Experiments with Platinum and Silver.—Silver is the best-known conductor of electricity. Platinum, on the other hand, offers far more resistance than silver to the passage of the electric current.

(1.) The consequence is, that when the poles of a battery of, say, three Grove's cells are connected by a short piece of silver wire, the wire is comparatively cool, the heat being confined to the interior of the battery.

(2.) But let the silver wire be replaced by a platinum wire of the same diameter and length, and the heat generated in the platinum will be found to be ten times greater, because this metal offers ten times more resistance to the passage of the current.

(3.) It is a pretty experiment to send a strong voltaic current through a chain composed of alternate links of platinum and of silver wire. The same quantity of electricity will pass through each link, yet the platinum links will be made white-hot, while the silver links will remain so cool that they may be touched with the hand.

(4.) As we decrease the *length* of a wire we *lessen* its resistance. This may now be shown by sending the same strong current first through a *long* piece of platinum wire, which will offer so much resistance that little electricity can pass, and consequently little heat can be developed; then through a *shorter* piece of the same wire, which

HEATING EFFECTS OF THE CURRENT. 83

will be made to glow brightly; and lastly, through a very short piece, when the heat produced will probably be so great as to *melt* the platinum, although to do this requires a temperature of 3,080° F. By electricity we can, in fact, obtain a higher temperature than in any other way, and by its means every known substance can be melted with the exception of *carbon*.

(5.) If, while a certain length of the platinum wire is red-hot, we dip one part of the hot wire into cold water, then the resistance of the wire will be diminished, *more* electricity will pass, and the other portion of the wire will become *white-hot*.

(6.) When a powerful current is sent through a coil of platinum wire placed inside a kettle, the water may be made to boil in a very few minutes by the heat developed by the electrical resistance of the platinum. In all these cases we have, in fact, examples of the conversion of electricity into heat.

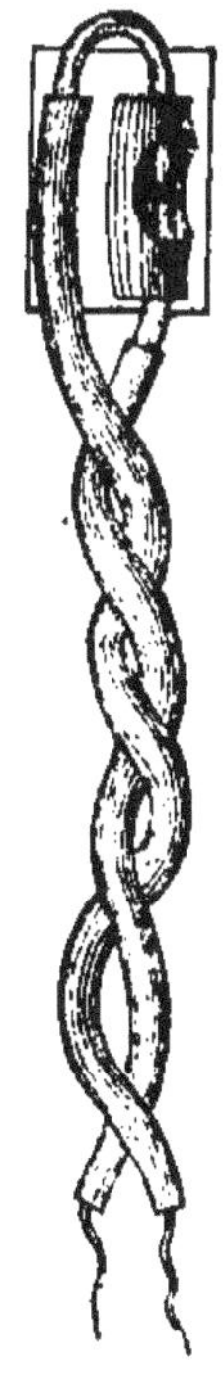

85. Blasting by Electricity.—The heating of platinum by the electric current enables us to explode torpedoes and to fire blasts and mines. The current passes from the battery through thick insulated copper wires to a *fuze*, which is placed in the midst of the gunpowder or other explosive substance of which the blasting charge is composed. Inside the fuze (Fig. 41), a short piece of fine platinum wire completes the circuit.

Fig. 41.—Fuze for blasting.

When the current is allowed to flow, the platinum becomes white-hot and ignites the fine-grained gunpowder with which the little wooden box is filled. In this way the blasting operations can be conducted from a distance, and therefore in perfect safety; while charges of gun-cotton (torpedoes) can be exploded at the bottom of the sea by means of wires passing from them to a battery placed in a boat or upon the shore.

XIV.—LUMINOUS EFFECTS OF THE CURRENT.

86. The Electric Spark produced by a Voltaic Current—87. The Voltaic Arc—88. The Incandescent Light—89. Vacuum Tubes.

86. The Electric Spark produced by a Voltaic Current.—The electricity produced by friction possesses great electro-motive force, so that the spark from a plate machine will usually leap across a space of two or three inches, overcoming the resistance of the air. On the other hand, the electro-motive force of a voltaic cell is much smaller, so that when we bring the ends of the wires connected with the poles of even a large battery—of, say, forty Grove's cells—within the one-hundredth part of an inch of each other, the voltaic current is wholly unable to leap across even this small interval. It has, indeed, been calculated that to produce a one-inch spark in air would require a difference of potential exceeding that furnished by 70,000 Daniell's cells! The most powerful battery ever constructed belonged to Mr. De la Rue, and was composed of 11,000 cells; yet the spark it could furnish was only one-seventh of an inch in length. But if, after closing the circuit, we suddenly separate the ends of the wires, a small spark will be visible, although we may only be using one or two cells. This spark is due to a momentary direct induced current, which flows in the wire itself, and which has a high electro-motive force. Fastening one wire from a battery to a *file*, and drawing the end of the other wire along the face of the file, a stream of sparks can be obtained.

The long sparks produced by induced currents will be described (see chapter xvi.) in connection with the induction coil.

87. The Voltaic Arc.—Connect the wires proceeding from the poles of a battery of forty or fifty Grove's or Bunsen's cells with two pointed pieces of carbon. When the pieces of carbon are made to touch, and are then separated half-an-inch or so, a kind of electric flame, called the *voltaic arc* (Fig. 42), is produced between the two carbon points. The carbon points become white-hot, and some of the carbon is carried across, in the state of vapour, from point to point, thereby completing the circuit. The light is so intense that it is impossible to look at it without pain (Fig. 43); so that when we desire to use these *arc-lights* for illumination, they are enclosed in ground-glass globes, by which the intense glare of the light is softened and diffused. With

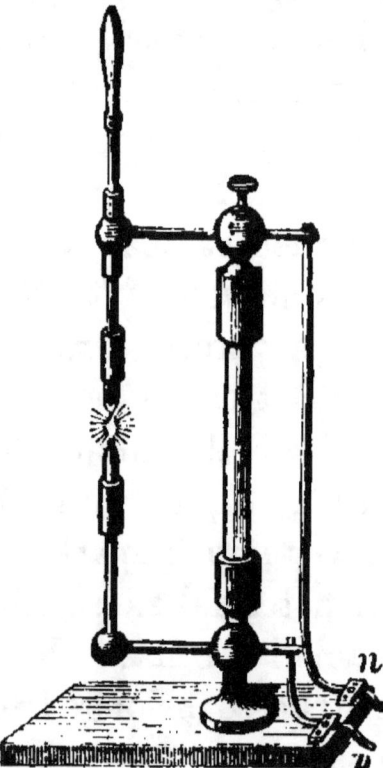

FIG. 42.—Mode of producing the voltaic arc.

the powerful current from a large dynamo machine worked by a steam-engine, a light exceeding that of 100,000 candles has been obtained from a single arc-light.

88. The Incandescent Light.—The light of the electric arc is suitable for illuminating the streets, and such large spaces as railway stations, the rooms of factories, etc., but for general use in our houses it is not well adapted. For lights of small intensity, not exceeding 100 candle power, it is far simpler

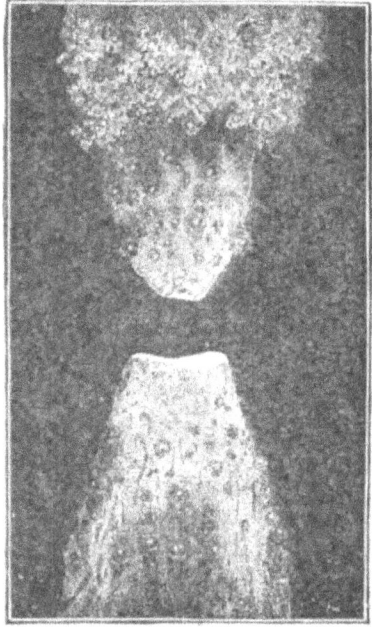

FIG. 43.—Carbon poles of voltaic arc.

and cheaper to use the electric current to raise to a white heat a thread of some infusible conductor. Platinum wires have been tried; but fine threads of carbon are much better, and are now universally employed. The carbon thread must be enclosed in a little glass globe exhausted of air, so that it may not burn away. Such incandescent lamps have been invented by Swan (Fig. 44), Edison, and others. They give a pure, white light which enables us to distinguish *colours* as well as by daylight, which does

FIG. 44.—Incandescent lamp.

not *heat* the air, and which does not render impure the air of our rooms.

88 LUMINOUS EFFECTS OF THE CURRENT.

89. Vacuum Tubes.—Under ordinary conditions, air offers much resistance to the passage of electricity. But when the greater part of the air has,

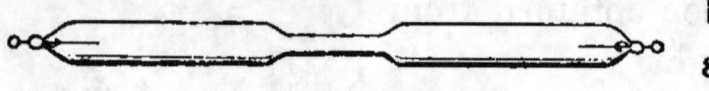

FIG. 45.—Vacuum Tube.

by means of an air-pump, been removed from the interior of a glass tube or vessel, then the electricity is enabled to pass with comparative ease. Tubes which have been so prepared are known as *vacuum tubes* (Fig. 45), and they exhibit a beautiful glow of light when the platinum wires, or electrodes, which are fused through the glass ends of the tubes, are connected with an induction coil. The light inside the tube varies with the gas contained in the tube. Carbonic acid gas gives a *white* light, nitrogen a *rosy* colour, and hydrogen *bluish* (but crimson where the tube is narrow). When the vacuum tube is made of an oval shape, it is termed the electric egg (Fig. 46). There is usually a violet glow of light round the negative electrode. If the exhaustion is carried too far, so that all or nearly all the air or other gas has been removed, then the electricity cannot pass, and there is

FIG. 46.—Electric Egg.

no light. On the other hand, when air is freely admitted to the tube the ordinary electric spark is seen just as in the air outside.

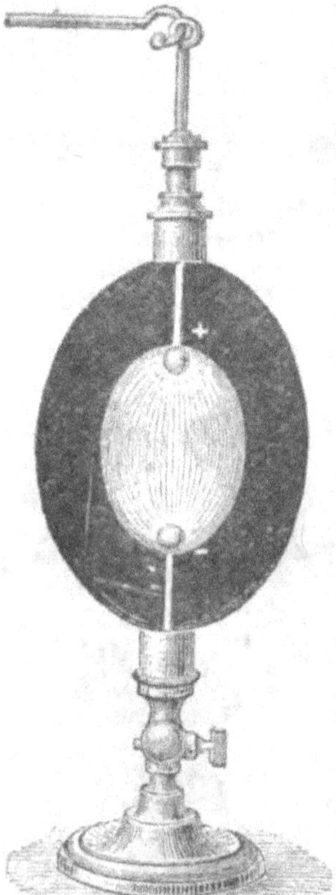

XV.—INDUCED CURRENTS.

90. Nature of Induction—91. Inductive Action requires a Medium—92. The Molecular Vibration Theory of Electricity—93. Induced Currents produced by a Magnet—94. Induced Currents produced by Currents—95. Table of Induced Currents—96. Magneto-electric Machine—97. The Dynamo Machine.

90. Nature of Induction.—By *induction* we mean *action at a distance.* Any one who has experimented with a magnet will remember how the steel bar caused needles to jump to it from a distance of an inch or more, and how one magnet can attract or repel another (suspended) magnet from a distance of several inches; these are examples of *magnetic* induction. In the same way, bodies which have received a charge of electricity by friction act upon other bodies *at a distance.* Light substances—bits of paper, feathers, etc.—rise from the table and rush upwards when excited rods of glass or of sealing-wax are held over them. We shall find exactly the same thing in voltaic electricity; a voltaic current can *induce* a current in a neighbouring conductor. Another proof of the connection between magnetism and electricity will also be found in the fact that a current can be produced in a wire by the approach or by the withdrawal of a magnet (Fig. 47).

91. Inductive Action requires a Medium.—If we attempt to explain *how* it is that the magnetic or electric force can be transmitted in this way, we find ourselves unable, as yet, to fully answer the question. It seems certain that there must be some substance, some *medium*, between any two bodies, if force is to be transmitted from the one to the other. But what is the medium by which electric force is transmitted? It is not the *air*, for the force acts equally well through a *vacuum*—a space devoid of air. Possibly the medium may be found in the *ether*, the rare fluid which we believe to fill all the space between the Earth and the Sun and stars.

Fig. 47.—Induced current produced in wire by magnet. *a*, coil into which magnet is introduced; *b*, coil surrounding a magnetic needle.

92. The Molecular Vibration Theory of Electricity.—The most probable theory of electricity is that which considers it to be—like heat and light—a peculiar *vibration* of the molecules of the electrified bodies. Such vibrations would produce *waves* in the surrounding ether, and when these waves struck another body they would cause its molecules to vibrate in the same way as those of the first body; and thus the second body would be electrified by the *influence* or *inductive action* of the first. Experiments performed with vibrating discs, by which air-waves are produced, strongly support this theory; for it is found that such air-waves can be transmitted to other discs, which are thereby set in motion.

INDUCED CURRENTS. 91

93. Induced Currents produced by a Magnet.—The space around a magnet over which its magnetic power extends is known as the *magnetic field*. This may be beautifully shown by placing a card upon a magnet, and dusting iron filings over the card. The filings arrange themselves in curves, spreading out from either pole, which curves are known as *lines of force*. In 1832, Faraday discovered that when a conductor, forming a closed circuit (as a hoop of copper wire), was moved *across* these lines of force, an electric current was produced in the conductor.

Fig. 48 shows a coil or *helix* of copper wire, the

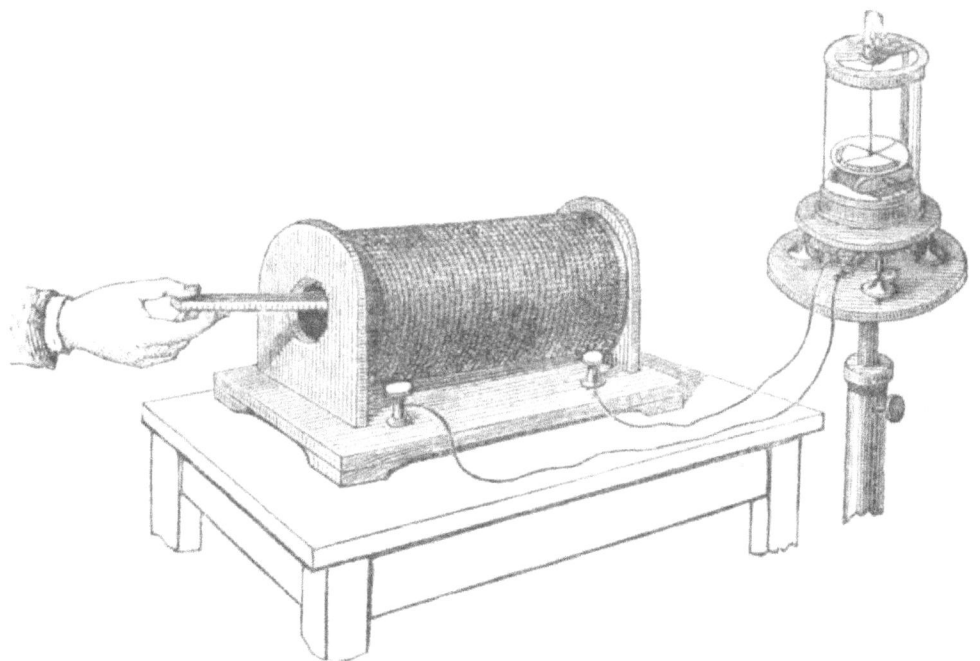

FIG. 48.—Wire helix connected with galvanometer: magnet inserted.

two ends of which are connected with a galvanometer. When a magnet is suddenly placed within the coil, a momentary *inverse* current traverses the wire and deflects the needle. While the magnet remains at rest near or within the coil, it induces no current; but let the magnet now be suddenly removed, and a *direct* induced current will be set up in the wire, and the needle will again be deflected, but in the opposite direction.

The strength of the induced current varies with the strength of the inducing magnet. If a strong electro-magnet be used instead of a steel magnet, the induced current will be strong in proportion. It is, of course, immaterial whether the magnet be brought to the wire, or whether the wire be moved towards the magnet or across its lines of force. An induced current will be produced in each case.

94. Induced Currents produced by Currents.—If the magnet be now replaced by a wire through which a strong voltaic current is flowing, we shall be able to show that the approach or withdrawal of the wire carrying this current will produce momentary (induced) currents

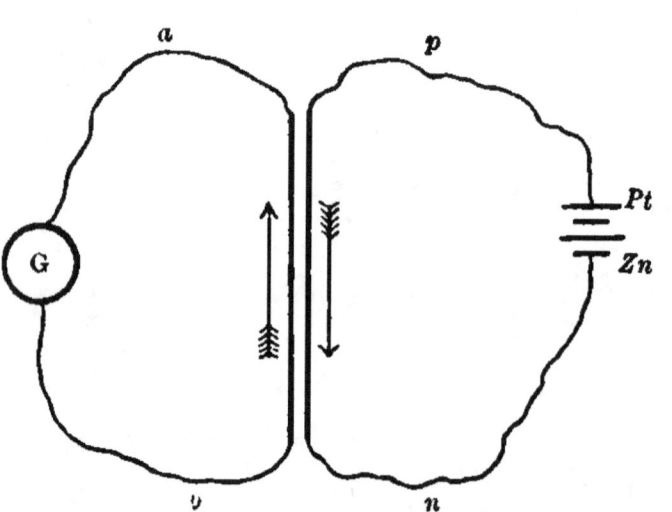

FIG. 49.—Induced current produced by current.

in a closed circuit. Let $p\,n$ (Fig. 49) be the primary wire, connected with a voltaic battery, along which a current is flowing from p to n; and let $a\,b$ be the secondary wire, connected with the galvanometer, G. When the primary wire is suddenly brought close to the secondary wire, a momentary *inverse* current flows in the latter from *b* to *a*, and declares its presence by deflecting the needle of G. When the primary wire is withdrawn, a momentary *direct* current flows along the secondary wire from *a* to *b*, and deflects the needle of G in the opposite direction. In order to produce stronger induced currents, it is usual to make both the primary and the secondary wires into large coils, so that the former may be placed within the latter, just as the magnet was placed within a similar coil.

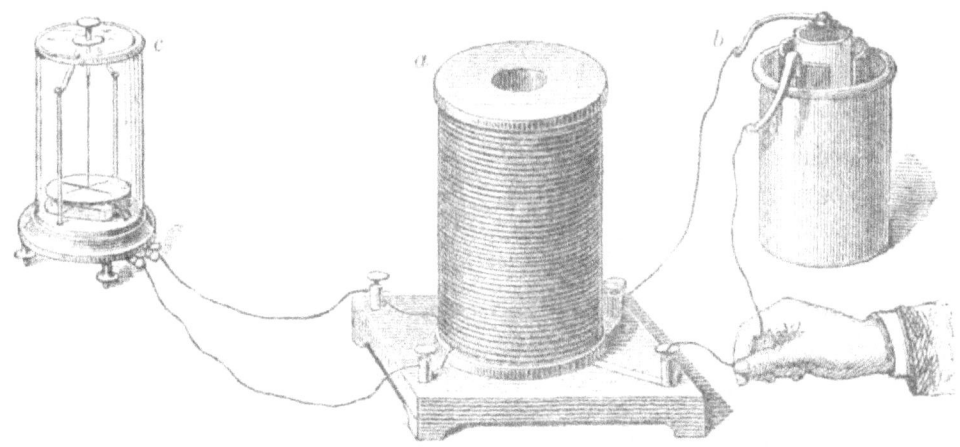

Fig. 50.—Induced currents produced by currents. *a*, Double coil of copper wire, the two ends of the outer coil being connected with the galvanometer, *c*, while the two ends of the inner coil are connected with the Bunsen cell, *b*.

Completing or *making* the battery circuit while the primary wire lies near or within the secondary, produces exactly the same effect as bringing it suddenly near. Breaking the primary circuit produces the same effect as withdrawing the primary (Fig. 50). While the primary current flows *steadily*, and both the wires remain *at rest*, no induced current passes. But an *increase* of strength of the primary current produces the same effect as its *approach*.

A *diminution* of strength of the primary current causes the same effect as its *withdrawal*.

95. Table of Induced Currents.—We may now sum up, in the form of a table, what we have learned about the production of induced currents by magnets and by primary currents.

TABLE OF INDUCED CURRENTS.

	Momentary INVERSE Currents are induced in the Secondary Circuit.	Momentary DIRECT Currents are induced in the Secondary Circuit.
By means of a Magnet.	While *approaching*.	While *receding*.
By means of a Current.	While *approaching*, or *beginning*, or *increasing* in strength.	While *receding*, or *ending*, or *decreasing* in strength.

96. Magneto-electric Machines.—Many machines are now used in which bobbins of copper wire, containing a core of soft iron, are made to rotate in front of or between the poles of a magnet. Since a current is produced in the wire by the inductive action of the magnet, these machines are known as *magneto-electric* machines. Small machines of this nature are much used to produce electric currents which may be passed through the human body, and which are useful in certain diseases of the nervous system. These medical magneto-electric machines consist of a steel magnet, two bobbins of wire with a soft iron core, and the necessary wheels for rotating the bobbins, all enclosed in a wooden box. A wooden handle screws on, by which the wheels are set in motion; while two brass handles—one to be held in each hand—are connected by flexible wires with the ends of the wire which is wound upon the bobbins. One person then turns the handle, while another holds the brass terminals, and allows the induced current to pass through his body.

97. The Dynamo Machine.—Magneto-electric machines of large size, in which the steel magnet is replaced by powerful electro-magnets, S N, are now largely used to produce the electric light; they are known as *dynamos* (Fig. 51). The coils of wire of each dynamo are made to rotate by means of a steam-engine or a gas-engine.

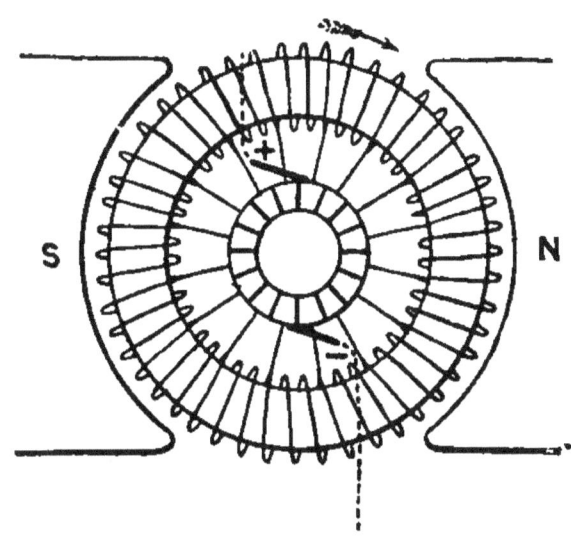

Fig. 51.—Dynamo machine: gramme ring armature.

XVI.—THE INDUCTION COIL.

98. Parts of the Induction Coil—99. Working of the Induction Coil—100. The Interrupter, or "Make and Break"—101. The Commutator, or Current Reverser—102. The Condenser—103. Large Induction Coil.

98. Parts of the Induction Coil.—Induced currents have a very high electro-motive force, resembling in this respect the charges of electricity obtained by friction. This high E.M.F. enables induced currents to overcome resistances which the primary current from any voltaic battery would be wholly unable to pass through. The instrument most frequently used to examine the effects of induced currents is known as the induction coil; it is frequently called Ruhmkorff's coil, after an electrician by whom it was greatly improved.

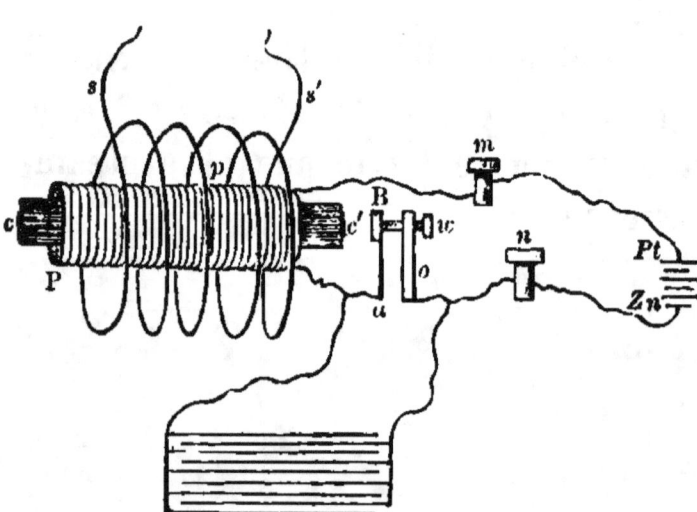

FIG. 52.—Induction Coil with condenser.

The parts of an induction coil shown in Fig. 52 include a core, $c\ c'$, formed of a bundle of iron wires, round which are coiled three or four layers of well-insulated stout copper wire, P p, to form the primary. The ends of this wire pass to the binding-screws, $m\ n$, to which are also fastened the wires from the battery, $Pt\ Zn$. A tube of thin ebonite surrounds the primary wire, insulating it and separating it from the thin secondary wire, $s\ s'$, of which only a few coils are shown; but this secondary wire is of great length, several miles of it being frequently used for a single coil. The ends of the secondary wire are shown at $s\ s'$. They are usually fastened to sliding brass rods supported by glass pillars, which, with the other parts of the coil, are supported by a mahogany base.

99. Working of the Induction Coil.—When the primary wire is connected with the battery, it is traversed by a voltaic current, by whose action the core is magnetized. By the joint action of core and primary an induced *inverse* current is made to traverse the secondary wire; and this induced current has so much electro-motive force that it leaps from s to s', producing a small spark. When the primary circuit is broken, a much more powerful *direct* current is induced in the secondary wire, and the length of the spark produced by the passage of this direct current from s' to s may be many inches. In ordinary coils it is the induced *direct* current only which produces the sparks; it is much more intense than the induced *inverse* current.

100. The Interrupter, or "Make and Break."—To save the trouble of interrupting and completing the battery circuit, and also to do this more rapidly than could be done by hand, the contrivance called an "interrupter," or "make and break," is placed in the circuit at B. This consists of a steel spring, *a* B, having a circular piece of iron attached to B, and another upright piece of iron, *o*, through the top of which passes a small iron screw, *w*. When the battery is not connected with the binding-screws, *m n*, the screw, *w*, touches the spring, B. But when the current flows and the core, *c c*, is magnetized, it attracts B and draws it away from *w*, by which means the circuit is *broken*, and the flow of the primary current ceases. But the core then also ceases to be a magnet, and therefore the spring brings B back again into contact with *w*, and again the primary current passes. Thus the induction coil is made complete and *self-acting*. The form of interrupter here described is also known as a *vibrator* and as a *contact-breaker*.

101. The Commutator, or Current Reverser.—Having provided for the *continuous* working of an induction coil, it is also desirable to be able instantly to throw the coil into action, or to stop it, or to reverse the direction of the battery current. These objects are readily effected by means of a *commutator* (Fig. 53), which consists of a barrel-shaped piece of ivory or ebonite provided with a handle, and having two brass cheeks which are connected with the battery poles. This ebonite barrel stands on the wooden base of the coil, and has an upright brass spring on each side of it, these brass springs being connected with the ends of the primary coil. When the barrel is turned so that its brass cheeks touch the brass springs, then the circuit is complete and the coil begins to work. When a quarter-turn is given to the commutator, the circuit is interrupted and the coil ceases to work. But when the barrel is turned half-way round, the current is reversed, or sent in the opposite direction round the primary coil.

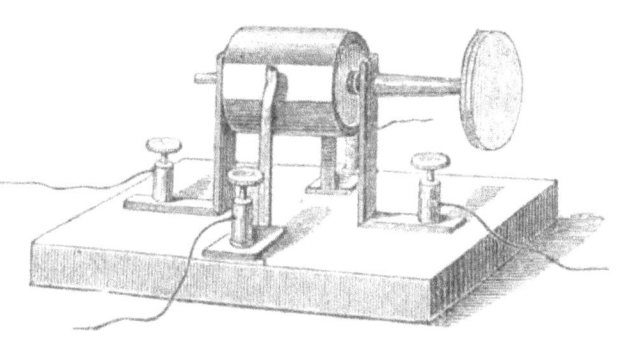

FIG. 58.—Commutator. Springs in contact with brass cheeks; current flowing.

If it were not for this contrivance, we should have to join, remove, or change the position of the battery wires leading to the binding-screws of the coil every time that we wished to start, or stop, or reverse the direction of the current.

102. The Condenser.—Lying within the wooden base of most coils (which base is made hollow like a box to contain it) there is a small *condenser*, formed of alternate sheets of tinfoil and waxed paper. This condenser is connected with the ends of the primary wire, and is so arranged that the current flows into it whenever the primary circuit is broken. The result is (1) that the break of circuit is made more *sudden*, and (2) that when contact is again made the stored-up electricity in the condenser causes the fresh current to attain its full strength *gradually*, thereby producing a weaker inverse current in the secondary wire. For in the practical working of a Ruhmkorff's coil, it is found best to use the *direct* induced current only, and to weaken, or suppress altogether if possible, the *inverse* current in the secondary wire. Much longer sparks are thus obtained.

103. Large Induction Coil.—The largest coil yet made was constructed for the late Mr. Spottiswoode. Its weight was 15 cwt., its length 4 feet; the core was a bundle of iron wires 44 inches long by 3½ thick, weighing 67 lbs.; the length of the primary copper wire was 660 yards, and of the secondary wire no less than 280 miles. With a battery of thirty Grove's cells (quarts) this giant coil gave sparks of 42½ inches in length, which resembled small flashes of lightning. A diagram of a large induction coil is shown in Fig. 54.

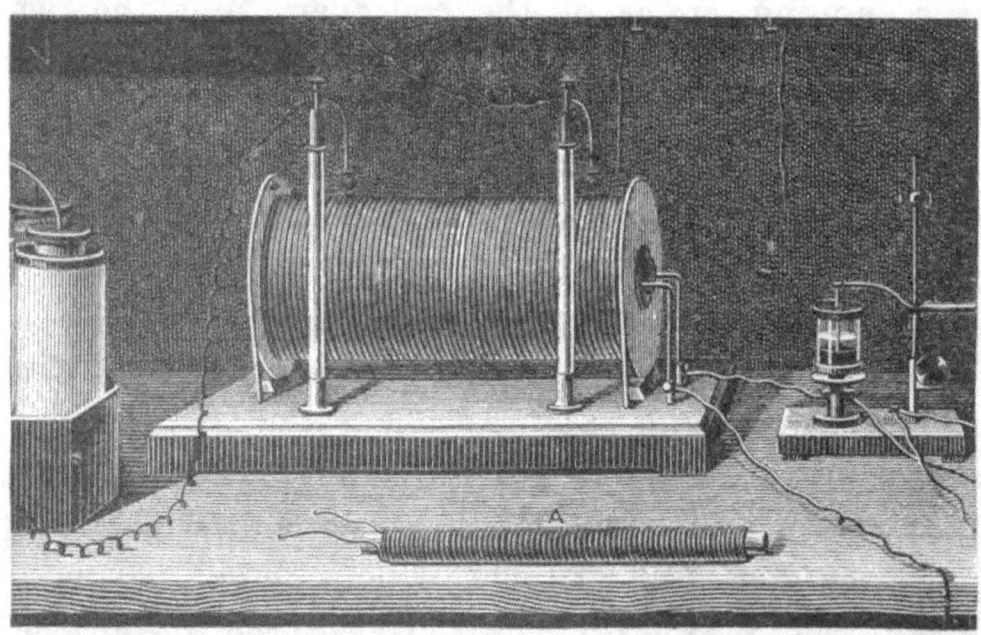

Fig. 54.—Ruhmkorff's Coil. A, Core detached. The battery by which the coil is worked is shown (in part) on the left. On the right is a new form of "make and break," consisting of a glass vessel containing mercury, into which a spring just dips.

APPENDIX.

EDUCATION DEPARTMENT.

Syllabus for Magnetism and Electricity; Fourth Schedule, Specific Subject, No. XII.

THIRD STAGE—VOLTAIC ELECTRICITY.

Voltaic or chemical electricity. The voltaic battery and notions of a current.

Chemical effect of a current. Electrolysis.
Magnetic effect of a current. Galvanometer.
Induced currents.
Electro-magnets. The electric telegraph.

EXAMINATION QUESTIONS.

I.

1. By what other names is voltaic electricity known?
2. Explain the "two-fluid theory" of electricity. By what signs do you represent the two fluids?
3. Describe clearly one experiment proving that electricity is developed by the *touching* of two metals.
4. How could you show that *voltaic* electricity is of the same nature as the electricity produced by *friction?*

II.

5. What happens when a strip of ordinary zinc is placed in acid water? How could you prevent the acid from dissolving the zinc?
6. Describe, and give a diagram of, a simple voltaic cell.

III.

7. How many chemical *elements* are there? Name those most frequently used in voltaic electricity, and give their symbols.

APPENDIX.

8. What *compounds* are largely used in voltaic electricity? Of what elements is each of these compounds composed?

9. Explain the terms *body, particle, molecule,* and *atom.*

IV.

10. Explain fully the chemical action in the simple cell by which a current of electricity is produced.

11. How many currents flow out of a simple cell? Which of these currents do we disregard?

12. What is meant by a *closed circuit?* What happens when the circuit is *broken?*

V.

13. Explain clearly why the current from a simple cell decreases rapidly in strength.

14. Describe the construction of Smee's cell. What is the object of covering the silver plate with a deposit of platinum?

15. Which do you consider the best *one-fluid* cell? Give a reason for your answer.

VI.

16. Describe, and give a diagram of, a Daniell's cell. What is the use of the copper sulphate?

17. State all the chemical actions which take place in a Grove's cell. Why could we not use copper instead of platinum in this cell?

18. How does Bunsen's cell differ from Grove's? Why is it cheaper?

VII.

19. Explain the word *potential.* If two bodies of the *same* electrical potential be connected by a conductor, what happens?

20. Arrange the following substances in order, putting first those bodies which offer *most* resistance to the passage of the electric current—pure water, mercury, silver, platinum, copper, carbon, brine, air.

21. Of what is a voltaic battery composed? How would you connect three Daniell's cells if they were required for telegraphic purposes?

22. What is meant by the "internal resistance" of a battery? How can it be reduced?

VIII.

23. What instrument would you employ to show the decomposition of water? Give a drawing of it.

24. How is it that a single cell, however large, is unable to decompose water?

25. Describe exactly what happens when a current is sent through a voltameter containing a solution of copper sulphate. What difference

would it make if *mercury* were used instead of the copper sulphate solution?

26. How would you coat a copper spoon with silver? What name is applied to the process?

IX.

27. How could you cause the muscles of a dead animal to contract? Who first noticed this?

⁕28. Who was Volta? What explanation did he give of the muscular movements referred to in the last question?

29. I give you a penny, a half-crown, a piece of wire, and a cup of salt water, and require you to produce a voltaic current; how would you do it?

X.

30. What contrivance is commonly used to detect the existence of a voltaic current in a wire? Give a sketch.

31. Write out Ampère's rule.

32. What instrument would you use to detect *very feeble* currents of electricity? How could you arrange *two* magnetic needles so that a current could deflect the pair more easily than a *single* needle?

33. Explain the words *insulate, fluid, magnetic field, deflect, galvanometer*.

XI.

34. How would you convert a kitchen poker into a powerful magnet? Would its magnetic power be temporary or permanent?

35. Is it possible to cause a piece of copper to exhibit the properties of a magnet? If so, how is it to be done?

36. For what special purposes are electro-magnets employed?

XII.

37. When was the electric telegraph first used in England? Upon what discovery does its action depend?

38. Give a sketch of the arrangement of battery, key, line-wire, galvanometer, and earth-plates by which a telegraphic message can be sent from any one place to any other. What is the use of the *key* or *tapper?*

39. If a current was flowing from York to London, how would it affect a magnetic needle placed underneath the wire?

XIII.

40. A current from a battery of ten Grove's cells is sent first through a copper wire, and then through a platinum wire of equal length and thickness. Describe and account for the effect in each case.

41. A thermometer placed within a cell of a battery showed the contents to be at a temperature of 50 degrees before the circuit was com-

www.ingramcontent.com/pod-product-compliance
Lightning Source LLC
Chambersburg PA
CBHW062355220526
45472CB00008B/1812